The Military and Social Research

Edited by
Working Group The Military and Social Research
and Chance Switzerland – Working Group for
Security Issues

Volume 48

Ina Wiesner

Importing the American Way of War?

Network-Centric Warfare in the UK and Germany

Nomos

Die Deutsche Nationalbibliothek lists this publication in the Deutsche Nationalbibliografie; detailed bibliographic data is available in the Internet at http://dnb.d-nb.de.

Zugl.: Florenz, European University Institute, Diss., 2011

ISBN 978-3-8487-0496-5

1. Edition 2013

Contents

Acknowledgements

This publication is based on a European University Institute (EUI) Ph.D. thesis. I am grateful to the EUI, the Arbeitskreis Militär und Sozialwissenschaften (AMS) and the NOMOS Verlag. For their comments and advice I would like to thank Pascal Vennesson, Gerhard Kümmel, Adrienne Héritier and Theo Farrell.

This book has been published with generous financial support from the European University Institute. I appreciate this very much.

Ina Wiesner Berlin, February 2013

1 Approaching the Diffusion of Network-Centric Warfare

This book examines the adoption of the network-centric warfare (NCW) concept by the British and the German Armed Forces at the beginning of the 21st century. NCW stands for the belief that through the extended application of information technologies and the integration of intelligence, reconnaissance, surveillance, command and control systems and weapons into one single network, military organisations will be able to better collect crucial military information, thus reducing operational uncertainty and gaining a decisive information advantage over their enemies. This would eventually translate into superiority over the enemy on the battlefield.

The concept had originally been developed in the U.S. in the late 1990s and a number of U.S. military allies and partners – among them the United Kingdom and Germany – later on started to adopt this U.S. innovation. However, in a number of cases the adopted national versions of NCW diverged considerably from the original U.S. concept. This book seeks to uncover the reasons for the variation in concept adoption.

NCW had its roots in the technological advances of the information revolution, starting in the 1970s. Conceptually, it grew from the promises some authors saw in the emergence of information and communication technologies for the business world and, eventually, for military operations (Toffler 1980; Toffler/Toffler 1993). "Cyberwar is Coming" was the title of one of the early articles that speculated about the changing nature of war against the background of the information revolution (Arquilla/Ronfeldt 1997). While initially rather focused on the mere exploitation of technological advances, U.S. defence experts, such as Andrew W. Marshall, eventually began to theorise about the impact of new technologies on U.S. military strategy and doctrine, which ultimately resulted in a debate about the Revolution in Military Affairs (RMA) during the 1990s (Adamsky 2010; Rosen 2010).

Operation Desert Storm (1991) provided the trigger for the U.S. Armed Forces' intensified digitisation efforts during the 1990s. Analysts claimed that during the campaign electronic warfare, especially radar technology, precision-guided munitions (PGM) and the Global Positioning System (GPS) had been the "decisive element" for the quick victory of the U.S. military over Iraqi troops (Eshel 1991: 38). Hence, after this successful demonstration of how the use of information technology can lead to a measurable military advantage, U.S. Armed Forces engaged in a major effort to digitise their ISR, C2 systems and weapon platforms. In the late 1990s, the emergence of the NCW concept added a new quality to these digitisation attempts. NCW entailed more as just merely improving *existing* ways of military engagement through the use of technology. NCW was rather understood as a completely *new* operational concept in which the network idea presented a total change in the way military operations – at the tactical and operational level – could be conducted.

NCW has been developed by "innovative troops and forward-thinking Pentagon 'mavericks'" (Scott/Hughes 2003), notably Admiral Arthur Cebrowski, a naval officer and President of the Naval War College, and John Garstka, a scientific and technical advisor for the Joint Chiefs of Staff Directorate for C4 Systems (command, control, communications, computers). Preceded and influenced by the idea of a 'system of systems' (Owens 1995, 1996, 2001), the network-centric concept appeared first in a publication in 1998 (Cebrowski/Garstka 1998). A year later Pentagon officials published a book entitled *Network-Centric Warfare* (Alberts et al. 1999). Today this volume is one of the most comprehensive and authoritative accounts of the concept. According to the authors (Alberts et al. 1999: 2), NCW is an "information superiority-enabled concept of operations that generates increased combat power by networking sensors, decision makers, and shooters to achieve shared awareness, increased speed of command, higher tempo of operations, greater lethality, increased survivability, and a degree of self-synchronization. In essence, NCW translates information superiority into combat power by effectively linking knowledgeable entities in the battlespace."

In 2001, then U.S. Secretary of Defense Rumsfeld made NCW a core feature of the U.S. military transformation. U.S. Department of Defense thinking purported that "the network will be the single most important contributor to combat power" (U.S. Department of Defense 2001). Since then, NCW has been implemented by the U.S. Armed Forces and applied in combat to such an extent that policy makers and observers even claimed the concept had thus been validated. (Lawlor 2005; Rumsfeld 2002a, b)

As a consequence of this apparent success story, other states became interested in the net-centric concept, too. Since 2001, the United Kingdom (UK), Germany, Australia, Denmark, Sweden, Italy, the Netherlands and France have introduced some form of NCW. Even the North Atlantic Treaty Organization (NATO), and lately the European Union (EU), have committed themselves to the network idea. These states have officially declared a desire to introduce NCW, or, in the case of NATO and the EU, to foster NCW in their member states. As a result, national military organisations committed to NCW have started to purchase new military equipment, set up new organisational entities and have added new processes, such as focusing on interoperability between their single services and between them and other allies or partners.

Despite the apparent attention, NCW received surprisingly little systematic research exists on this spread of NCW. Quite evidently the spread of NCW represents a case of military *concept diffusion*. In the academic literature, diffusion is defined as the "social process by which an innovation spreads through a social system over time" (Webster 1971: 178). Often innovations – be they technological or conceptual in nature – change during the course of diffusion and in consequence it is not uncommon that the diffusion outcome (adoption) varies considerably from the original innovation once disseminated by the innovator. At least this seems to be true for the diffusion of NCW. The Swedish, British, German and also a number of other national variants of NCW differed from the U.S. original in their general ambition, notably the extent to which NCW was introduced. They also differed substantively, i.e., concerning those aspects of NCW that were emphasised or, on the other hand, downplayed. The most visible give-away of the considerable differences in the national variants of NCW is the plethora of national acronyms created for NCW. Sweden's networking project was called Network-Based Defence (NBD); the Danish concept was known as Network-Based Operations (NBO); the UK, NATO and the EU call their initiatives Network-Enabled Capabilities (NEC), and Germany had even chosen a German term (Vernetzte Operationsführung, NetOpFü).[1]

There is also a temporal dimension in the adoption differences. Some states, notably Sweden and Australia, decided to adopt NCW early on and proceeded quickly with its introduction. Other states were comparatively late and slower in adopting NCW, exemplified by the special case of Austria, which has not yet overcome the phase of drafting an own NCW concept since 2004.

1 For official purposes the English term Network-Enabled Capabilities (NEC) had been proposed as a translation for the German NetOpFü. *NEC* will thus be used throughout the book to refer to the German networking project, too.

Generally, variances in concept adoption are not extraordinary events. But still, when it comes to the introduction of NCW it is puzzling why two comparatively similar military organizations, as Her Majesty's Armed Forces (hereafter referred to as the British Armed Forces) and the German Bundeswehr, differed considerably in their adoption of NCW: The UK was rather quick in its decision to introduce NCW and pursued the implementation of its own Network-Enabled Capabilities (NEC) concept forcefully, while the Bundeswehr took years to decide to have an own version of NCW and then implemented its concept of Vernetzte Operationsführung (NetOpFü) only slowly. The aim of this book is to account for the striking differences of NCW concept adoption in the UK and Germany in the period between 2001 and 2010.

The remainder of this first chapter entails an account of the analysis of NCW adoption and summarises the main argument advanced throughout the book. The chapter continues with a discussion of book's relevance for policy makers and the contribution it seeks to make in the literature on military change. A brief outline of the book's chapter structure is offered in the end.

1.1 Research Questions and Argument

Throughout the book, two questions will be central. First, how did the British and German military organisations differ in their adoption of the NCW concept, both in terms of adoption process and adoption outcome? Second, which explanatory factors can account for the differences?

To answer the first question three main categories had been identified which make comparisons of adoption processes possible. These are, first, the *timing and pace* of the NCW concept adoption in the UK and Germany, second, *concept faithfulness* (the degree to which the adopted concept resembled the original one); and, third, *implementation faithfulness* (the degree to which the new concept was implemented). Extensive research revealed that in the case of NCW adoption the UK and Germany differed considerably in each of the three categories of concept adoption.

Table 1: NCW Adoption Differences between the UK and Germany

	UK	Germany
Timing/Pace	Early/Quick	Late/Slow
Concept Faithfulness	Low	High
Faithfulness of Implementation	Moderate to High	Low

The second question aimed at singling out the factors that can account for the observed differences of adoption processes and outcomes between the two militaries. The chief argument throughout the book is that military organisations introduce foreign concepts not only in order to increase their efficiency and military effectiveness, but also with a view towards increasing their *institutional legitimacy.* When organisations are in need to increase their institutional legitimacy they might respond to certain demands of their institutional environment and introduce new policies that do not necessarily result in better or more efficient performance. The observed adoption variances in Germany and the UK, it is argued in the book, stem from the different composition of these efficiency/effectiveness and legitimacy drivers in the two cases.

1.2 Efficiency, Effectiveness and Legitimacy as Drivers for Change

When explaining the outcomes of military change, researchers usually take on one of two perspectives. They argue either from a rationalist perspective that sees military change as an inescapable necessity to improve military performance, hence securing national survival (Resende-Santos 2007; Waltz 1979), or they explain how military change is moulded by military culture (Adamsky 2010; Farrell 2001; Kier 1997; Nagl 2005; Terriff 2006) and institutionalist mechanisms (Goldman 2003; Goldman/Ross 2003). Although compelling in one way or the other studies that focus on one explanatory factor only run the risk of coming at the expense of explanatory power. In order to understand the differences in NCW concept adoption it can be fruitful to bring arguments from those rational, institutional and cultural explanations together (Fearon/ Wendt 2002).

Table 2: Drivers for NCW Adoption

Level	Effectiveness/Efficiency Issues	Institutional Legitimacy Issues
International	• Achieving or maintaining interoperability with the U.S. • Interoperability in NATO • Other states acquiring NCW (neo-realist argument) • Operational requirements	• Appeasing the hegemon (for example burden sharing) • Appeasing the alliance/ coalition (for example burden sharing) • Gaining international credibility • International prestige
National (Societal)	• Better protection of citizens, territory and state function (ends-oriented)	• Legitimate protection of citizens, territory and state function (means-oriented)
National (Political)	• Achieving strategic objectives • Better resource management	• Meeting budget priorities • Strengthening defence industrial base/Appeasing defence industrial lobby • Civil-military relations
Military Organisational	• Costs, casualties, operational and tactical effectiveness	• Inter-service relations, intra-service relations, military culture

In the case of NCW, the observed variance in concept adoption should not be explained by either rationalist or cultural accounts but – as table 2 exemplifies – by examining the different *composition*s of either rational drivers that are concerned with the increase of organisational efficiency or effectiveness and institutional drivers that are concerned with increasing the organisation's institutional legitimacy.

A causal relationship exists between the militaries' responsiveness to these environmental incentives and constraints to change. However, although the efficiency/legitimacy divide would offer an elegant dichotomy by which to categorise different cases of military innovation, both sorts of drivers are likely to be present in any case of military change. It is rather unlikely to find cases in which military concept adoption was driven purely by considerations of efficiency and effectiveness. Likewise sole strive for legitimacy is improbable, too.

Opportunities or pressures from the institutional environment are necessary for change. But at the same time they are not sufficient to account for differences in concept adoption. Internal factors, especially military culture can also foster military innovation and change (Adamsky 2010). The potential explanatory power of the cultural argument should neither be neglected nor discarded and attention will be paid to internal motivations throughout the book.

1.3 A Word on Methodology

This book provides an in-depth account of the introduction of NCW in the British and German militaries. Contrary to quantitative social science research in which only a few factors in a large number of cases are assessed an in-depth engagement with one, two or three country studies allow for a better understanding of the specifics of each case (Della Porta/Keating 2008). In the social sciences when more than one case is assessed with the help of a framework of categories one speaks of *comparative case study design* and specifically of structured focused comparison (George/Bennett 2004: ch. 3). Structured focused comparison suggests that cases – even though their uniqueness is acknowledged – can be analytically compared by using the same set of standardized questions. When conducting empirical research on the introduction of NCW in the UK and Germany the same research framework (i.e. the same set of questions) was applied to assess both cases. A look at the sub-headlines of the case study chapters (4 and 5) in the book reveals that they follow a similar "story line".

In the book the adoptions of NCW in the UK and Germany are presented in the form of linear temporal stories. Starting when the British and German Armed Forces became aware of NCW, each story continues with observations about the development of a national "version" of NCW and, ultimately, ends with the implementation of the NCW concept. In a way, the book *reconstructs* temporal lines of NCW concept adoption. It is recreating the order of events on paper. In the social sciences this method is often coined *process tracing*. This "important, perhaps indispensable, element of case study research" (Vennesson 2008: 224) enables social science researchers to systematically shed light on the black box that lies between the initial conditions and political outcomes in complex political processes (Checkel 2006, 2008; George/Bennett 2004: ch. 3).

The book looks closely on the NCW adoption differences between the British and German military. Explorative research on a larger number of NCW cases had initially revealed that especially the UK and Germany differed considerably in their NCW adoption outcome. This was puzzling as both militaries show a number of similarities with regards to size, capability and mission: Compared to the rest of so called NATO Europe, the UK and Germany both have quite powerful military organisations. Together with France, they form the 'big three' in Europe.[2] They might not be similar but they are comparable with a view to their military power – both quantitatively and qualitatively (Biddle 2004; Huntington 1957; Mearsheimer 2003). Comparisons of defence expenditures and military manpower in NATO-Europe show that both countries are closely positioned in the respective rankings. Both states, furthermore, have a strong defence industrial base and range among the top-ten weapon suppliers worldwide (International Institute for Strategic Studies 2010).

2 Currently, the 'big three' is a common journalist and academic expression for France, the UK and Germany in the context of European security. During World War II, however, it was used for the United States, with Britain and the Soviet Union as the allied coalition against Germany, Italy and Japan.

The security of both states is embedded in the transatlantic alliance. The UK and Germany, furthermore, have close bilateral military relations to the U.S. – Britain enjoys a special relationship with the U.S. through the close cultural linkages between both countries (Wallace/Phillips 2009). Similarly, the relatively young history of the German Federal Republic has been closely linked to U.S. military engagement – first, during WWII, against the Third Reich, and later, during the Cold War, as a security provider against the Warsaw Pact. Both states were actively providing security on the European continent after the Cold War, especially on the Balkans. Both supported the military missions of the European Union. The two countries participated in the War against Terror, with the UK being engaged in Iraq (2003–2009) and in Afghanistan (since 2001) and Germany making a substantive contribution in Afghanistan, too (since 2002). Both have – to varying degrees – experienced the impact of Islamic terrorism within their own borders.[3] Finally, both military organisations were subject to far-reaching military reforms (decreasing size of forces and re-structuring of forces) during the time of NCW introduction. Of course, apart from these similarities, the military organisations of the UK and Germany are dissimilar in many other respects, such a force posture, tradition, and military culture – however, so would be any other military organisation.

1.4 Sources Used for the Book

For the book about 50 pre-structured in-depth expert interviews with German and British military officials had been conducted (Weiss 1994: 9f.) Furthermore, official documents had been consulted, speeches and other oral or written contributions by actors directly involved in the processes of NCW adoption had been analysed. Of lesser importance had been newspaper articles and, finally, secondary sources. The book also benefitted from the author's personal experiences as a temporary member of the German Bundeswehr's *Concept Development & Experimentation* branch in the Transformation Centre (March–April 2010, November 2011).

1.5 Relevance and Structure of the Book

Studying the emergence and diffusion of NCW adds a little piece of understanding about how the information-age brought about changes in the military – a question first dealt with by the Tofflers (Toffler/Toffler 1993). Yet despite a vivid debate among military experts in the 1990s about how the military should exploit the advances in information technology, Johan Eriksson and Giampiero Giacomello made the argument that International Relations Theory only insufficiently addressed the impact of the Information Revolution on national and international security (Eriksson/Giacomello 2006). Thus, this book also contributes to a discussion about the more general impact the revolution in military affairs (RMA) had on British and German national security. It thereby reflects on the relation between military capabilities and military missions. Ideally, form follows functions and military organisations will be re-designed to fulfil the tasks assigned to them by the political level. Yet, this book shows, that to the contrary the or-

---

3 For the case of the London bombings in 2005 and the investigation thereafter, see http://news.bbc.co.uk/1/shared/spl/hi/uk/05/london_blasts/what_happened/html/. For the thwarted attempt to attack military and civilian targets in Germany, see http://www.swr.de/blog/sauerland-verfahren/, archive section (both accessed 26 February 2010).

ganisation of a military might affect the formulation of the national defence policy and that "[m]ilitary capabilities can influence decisions about the use of force in important ways" (Fordham 2004: 633).

Finally, this book contributes to the *theory of diffusion* in military organisations. Eliason and Goldman, who published one of the spare volumes on this topic explain that "despite the large body of scholarship on military innovation, remarkably few studies explore either historical or contemporary processes of diffusion of military innovations" (Eliason/Goldman 2003: 7). This book aims at filling this gap. It is concerned with the ways in which new foreign concepts are perceived, introduced, altered and implemented by military organisations. Consequently, this research tries to combine existing ideas of how best to understand the processes of military concept adoption, and hence military change and develop them further. Thereby, it contributes to discussions and debate in the fields of security studies but also – with a view to the theoretical argument – to organisational theory.

The following Chapter 2 introduces the NCW concept in greater detail. It shows how NCW evolved and had become the cornerstone of the U.S. military transformation between 2001 and 2006. Chapter 3 explains the theoretical framework that has been developed to assess NCW concept adoption in the UK and in Germany between 2001 and 2010. In the following chapters 4 and 5 both cases are discussed. The final chapter 6 contains a synthesis of the findings from the previous case studies and also discusses the wider implications of the results.

2 Network-Centric Warfare and U.S. Military Transformation

The network-centric warfare concept evolved in the late 1990s in the U.S. and considerably influenced the U.S. Armed Forces' transformation project (2001–2006). The group of authors that advocated NCW comprised of senior military officers and officials notably Arthur Cebrowski, Fred Stein, John Garstka and David Alberts. The emergence of the NCW concept can be dated back to 1998 when a seminal article by Cebrowski and Garstka appeared in the U.S. Naval Institute magazine *Proceedings* (Cebrowski/ Garstka 1998). The term Network-Centric Warfare made its first appearance in 1997 when the U.S. Navy expanded its digitisation programme Copernicus (later ForceNet) (Vickers/Martinage 2004: 53f.). A year later, Cebrowski and Garstka drafted the first comprehensive account of the NCW concept. In 1999, Alberts, Garstka and Stein authored a comprehensive volume in which NCW-theory was thoroughly developed and explained (Alberts et al. 1999). Although much has been written on NCW since the late 1990s, the first initial article, *Network-Centric Warfare: Its Origin and Future* by Cebrowski and Garstka (1998), the volume on *Network Centric Warfare* by Alberts, Garstka and Stein (1999), and also the book *Power to the Edge* (Alberts/Hayes 2003) are still the most comprehensive works on the subject.

The emergence of NCW in the literature only slightly preceded the U.S. military transformation project that was announced by the incoming Bush administration in 2001. The new concept embodied many of the beliefs held by military reformists in the Bush administration – first and foremost Secretary of Defense Donald Rumsfeld (2001–2006). NCW was based on information technology and (re-)discovered the importance of information in order to gain a decisive advantage over the enemy, thus arguing for a substitution of military mass for precision and attrition warfare for effects-based operations. Implicitly, NCW promised a way to further substitute manpower for technology.

NCW, it appears, was the right concept at the right time. With the beginning of the U.S. military transformation offering a well-timed window of opportunity, the support of powerful individuals from the executive branch and the combat operations in Afghanistan and Iraq – NCW quite quickly became an apparent success story. Not only was the concept made the cornerstone of U.S. military transformation, it became an *Exportschlager* – an important export item of the U.S. –, which other states, such as the UK, Germany, Sweden, the Netherlands and many more sought to introduce into their own military organisations as well.

The seeming success of NCW and the U.S. military transformation are largely intertwined. The following sections therefore introduce the NCW concept to the reader and place it into the wider context of the U.S. military transformation. Thereafter, the NCW programmes of the U.S. individual services (U.S. Army, U.S. Air Force, U.S. Navy) will be described and the application of NCW in war will be discussed. The chapter closes with an overview of the criticism that was raised in academic and professional military circles on NCW theory, its application in the wars in Afghanistan in 2001 and Iraq in 2003, and on the technology-prone U.S. military transformation in general.

A final word on terminology: Since 2003, the original term Network-Centric Warfare has been replaced more and more in professional circles with Network-Centric Operations (NCO). The explanation given by the original advocates of NCW-theory was that NCO would better mirror the fact that NCW-theory was applied to "a much broader domain of phenomena and [was] not limited to warfare" (Garstka 2003). However, the lukewarm reception of Network-Centric Warfare by European Allies might have helped to replace *warfare* with *operations*, especially when the lack of societal stabilisation instruments was revealed in the post-invasion phase in Iraq. As the name change was not accompanied by substantive changes to the concept, the original term NCW will be used throughout the book for the sake of coherence.

2.1 Defining Network-Centric Warfare

NCW is not a technology. It might be true that NCW relies heavily on information-technology (IT), but in essence, NCW is a *concept* about the networking of disparate and dispersed military entities. These entities can be electro-optical sensors attached to drones that circle over a specified territory to gather information; they can be so-called effectors or shooters, that is, weapon platforms or soldiers that hit a target; or, finally, they can comprise of command and control units, such as command posts in which decisions are made and military actions are directed. NCW-theory posits that, by linking these entities together into a networked structure, information between these units runs faster than in otherwise hierarchical military organisations. Furthermore, networking spans the existing boundaries between the various arms or military services (often referred to as stovepipes) and thus creates synergies that otherwise would not be exploited. Greater synergies and a faster flow of information, NCW theory finally posits, lead to information superiority over an enemy, and thus bring the decisive advantage that results in combat superiority.

Although the ‘father' of NCW, Arthur Cebrowski, avoided defining NCW – for him, “[n]etwork-centric warfare is a concept. As a concept, it cannot have a definition, because concepts and definitions are enemies. (…) if a concept can be defined it is no longer a concept" (Cebrowski 2003a: 16) – there are at least two possible perspectives on the purpose of NCW and, thus, definitions (Roxborough 2002: 71). According to NCW theory, it is about integrating relevant functional military entities (sensors, shooters, decision-makers and support) into a network allowing for new ways of fighting, for example swarming (Arquilla/Ronfeldt 2003). This is captured in the well-known definition of NCW provided by Alberts, Garstka and Stein (Alberts et al. 1999: 2): “We define NCW as an information superiority-enabled concept of operations that generates increased combat power by networking sensors, decision makers, and shooters to achieve shared awareness, increased speed of command, higher tempo of operations, greater lethality, increased survivability, and a degree of self-synchronization. In essence, NCW translates information superiority into combat power by effectively linking knowledgeable entities in the battlespace."

In practice, however, NCW highlights the importance of Command and Control (C2), or, more broadly, the area of Command, Control, Communications, Computers and Intelligence (C4I). From the second perspective, NCW is understood as a means of reducing the time needed to plan or to decide military action through information and communication technology and related adaptations in military organisations. Thus, NCW would improve established ways of fighting. In the beginning of the 21st century, NCW was chosen to guide military modernisation programmes in a number of states. The majority of acquisition projects targeted the realm of C4I and were comprised of new battle management systems or digital communication projects. Similarly, the majority of military service's efforts to implement NCW focused on the generation of a Common Operational Picture (COP).

In a given situation, a COP displays up-to-date, relevant information (depending on the situation, this may include geographic data, weather data or the position of own and enemy forces) not only to a command post, but also to military units in the field. In a networked environment, data can be fed into the COP in real-time, thus allowing the commander to react quickly to changing circumstances. An accurate COP ideally also displays coalition troops or troops form neighbouring command areas. Thus, a COP might limit the danger of friendly fire.

Related to the COP and more elaborate concepts include *shared situational awareness* and *shared situational understanding*, the former also incorporating a mutually shared perception of the situation (including its meaning), and, in the latter case, the addition of a mutual projection about future actions.

The ultimate aim of NCW is the self-synchronisation of forces in the theatre of war. Based on the COP (What is going on? Where are we and where are the others?), and knowledge about the command's intent (What is the desired effect?), units in the theatre of war should be able to readjust their plans and actions when circumstances change without necessarily having to wait for orders. For a number of reasons (technological, cultural and conceptual) self-synchronisation is difficult to achieve. Some even doubted if self-synchronisation is desirable at all, as it puts a question mark behind the responsibility and accountability of military action. Yet, the idea of self-synchronising forces seemed intriguing for the proponents of NCW.

In an attempt to grasp the application of NCW in combat, Arthur Cebrowski envisaged its usage as follows (quoted in Blaker 2007: 48f.): "The network centric model is different [from the hierarchical model, I.W.]. A command hierarchy exists, but it deals out 'command intent' allowing more discretion at the operating and tactical levels. In effect, decision power flows outward, 'to the edge' where combat units implement the command intent. Information flows horizontally, not through hierarchical chains, but through networks, and it is this robust network flow that allows a common understanding to emerge throughout the force quickly. That common understanding, coupled with well-trained personnel, versed in knowing to think about how what they are directly aware of can be affected by what is occurring beyond their direct cognizance, is what allows them to self-synchronize their actions with those of their co-protagonists. Bottom line: it will pay off in a leap in military effectiveness."

2.2 The Formulation of NCW

The way NCW was drafted departed from the usual way of formulating new military doctrines and concepts. First, NCW was not written by staff of a doctrine and development centre (such as the U.S. Army Training and Doctrine Command), but by a number of people, many of them at the end of their military careers. Second, NCW borrowed heavily from observations of the civilian world. Third, although the concept was pushed into the services by the relevant authorities soon after its formulation, it was still regarded as a *theory* by its authors.

Although the advantages of networking for military operations had been tentatively established during the 1980s, the authors explained their ideas concerning the virtues of networking by extensively referring to the business world. Subscribing to the argument made by Toffler and Toffler that the way a society makes wealth would influence the way it makes war (Toffler/Toffler 1993), the authors studied networked businesses (such as DELL or WalMart), and applied emerging principles, such as competitive awareness, virtual organisations, cost and risk suppression, precision manufacturing and focused logistics to combat situations (Alberts et al. 1999: 55ff.). Their observation of the business world was that information technologies would enable firms to create a high level of competitive awareness. Networking, furthermore, was enabling a variety of information-based relationships, which, in turn, also increased competitive awareness. As a result, time was being compressed and the tempo of their business operations increased. The ultimate outcome of networking and the use of information technologies in the business sphere was the domination of their competitive space by those companies that were networked (Alberts et al. 1999: 50ff.). In short, networking resulted in a competitive advantage.

The authors found that business operations and military operations were similar in a number of regards. Hence, they deduced a variety of implications for the military. One important example concerned battle-space entities. Applying the logic of outsourcing, the authors argued that, in a networked environment military sensors and shooters could become physically independent. In an undesirable "platform-centric operation" a military unit could only use information it collected with own sensors. In "network-centric operations" instead, a variety of sensors and shooters from different platforms could be linked together in a network so as to increase the input of information for a military unit but also the options of military actions (Alberts et al. 1999: 120).

Some argued later that the business metaphor was flawed. Frederick Kagan made the point that NCW was formulated at the height of the information boom but that during the economic downturn not all of the networked businesses survived (Kagan 2006: 256). Paul Bracken had argued earlier in a similar way and furthermore criticised NCW theory, after its formulation, for not being adapted so as to include lessons from the corporate disasters some of the revolutionary companies had glided into (Bracken 2002). Nonetheless, at the time of formulation, the business example and its translation to the military sphere appeared appealing.

A second important observation regarding the formulation of NCW is that the authors understood the concept as a theory, with several distinct readings of it being possible. Alberts, Garstka and Stein argued that the assumptions NCW makes about the causal relationships between physical networking, information dominance and, ultimately, combat superiority had to be validated by experimentation (Alberts et al. 1999: 218; Garstka 2000). With the set-up of the Network-Centric Operations Case Studies series – conducted with the participation of John Garstka – this perspective of NCW as a *theory* was put into practice. Also critical observers joined in the call for validation of NCW assumptions by arguing that NCW was a "theory desperately in need of doctrine" (Meiter 2006).

By emphasising the comprehensive character of the concept, Arthur Cebrowski gave another reason for why he considered NCW a theory (Cebrowski 2002): "Network centric warfare is not about technology per se – it is an emerging theory of war. That is, power comes from a different place, it is used in different ways, it achieves different effects then it did before. When that happens we call that a new theory of warfare. It is not about the network, rather it is about how wars are fought. How power is developed. During the industrial age, power came from mass. Now power tends to come from information, access and speed." Apparently, Cebrowski's understanding of *theory* departed from the deductive-nomological, constructivist or critical notions of theory that are prevalent in the social sciences (Mjøset 2001). This particular meaning of NCW as a theory could perhaps best be described as a *paradigm of warfare* (Biddle 2003).

Another way of understanding NCW as a theory is to acknowledge the normative underpinnings of the concept. James Blaker contends that Cebrowski developed a "New Theory of War" (Blaker 2007: 18ff.) backed by distinct assumptions concerning the nature of man, morality and conflict. In this view, network-centric warfare was morally superior to industrial-style attrition warfare, as it offered a way to discriminate between military targets and civilians (Arthur Cebrowski quoted in Blaker 2007: 31f.): "The moral use of the military force is not just a matter of its purpose. It involves how military force is used. Moral use of military force is parsimonious in its violence. It is driven by commitments to precision and discrimination. (…) Attrition strategies are inherently indiscriminate. Overwhelming force is inherently sloppy. Industrial war will harm the innocent. It carries, therefore an inherent immorality."

To conclude, the formulation of NCW departed from the 'usual' production of new military doctrines that focus on improving military effectiveness or that sought to solve a military problem by thinking in military terms and by finding military solutions – consider, for example, the development of the AirLand Battle Doctrine (Lock-Pullan 2005). Maybe because of this unusual creation of the concept it could mobilise initial support from within U.S. defence. The next sections elaborate on the details of the features and tenets of NCW.

2.3 Understanding NCW

At the core of NCW theory is a set of hypotheses about the interrelation between an information infrastructure (the network) on one side and effectiveness in combat on the other. The authors proposed that there was a logical chain – equated to the value chain in the business world – of culminating positive effects of networking in military operations. The components of this logical relation were summarized as the "Military as a Network-Centric Enterprise" (Alberts et al. 1999: 87ff.). After the initial propagation of NCW theory, conceptual work continued. In 2000, John Garstka theorised about domains of warfare and NCW (Garstka 2000). In 2003 the value chain and the domain model were merged into the Network-Centric Operations Conceptual Framework (NCO CF). The following sub-sections elaborate on the NCW Value Chain, the Domain Model and the NCO CF (Garstka 2003; Garstka/Alberts 2004).

2.3.1 The Network-Centric Enterprise (the NCW Value Chain)

The first building block of NCW is the Network-Centric Enterprise (hereafter the NCW value chain), which consists of culminated positive effects due to networking and has thus been compared to so-called value chains in the business world (Alberts et al. 1999: 29ff., 87ff.). The value chain, it is argued, is composed of a robustly networked force improving information sharing and collaboration. This enhances the quality of information and shared situational awareness. As a result, further collaboration and self-synchronisation is enabled and the speed of command is increased. Ultimately, this results in an increase in mission effectiveness.

The NCW value chain rests on the military functions of information gathering, information processing, decision-making, commanding, controlling and, finally, executing orders. It was inspired by Air Force Colonel John Boyd's theorising on the OODA-circle (observe, orient, decide, act) in 1974 (Angerman 2004). The first assumption of the chain is that a robustly networked force improves information sharing and collaboration. The key idea is that a network will span traditional platform-centric and stove-piped systems. Military action is platform-centric when "sensing and engagement capabilities reside on the same platform, and there is only limited capability for a weapons platform to engage a target based on awareness generated by other platforms" (Alberts et al. 1999: 95). Network-centric warfare, in contrast, aims at overcoming these limitations by allowing information flows between all battlefield entities regardless to which arm, service (or, coalition partner) they belong to.

Ideally, the network will also cross arms and service boundaries. A precondition for this is technical interoperability of the systems in use (computer, server, and so on). Interoperability is a condition in which information or services can be exchanged directly and satisfactorily between communication devices (U.S. Department of Defense 2008). Yet, interoperability is not easy to achieve. The services, often also different arms within one service, use different types of equipment and, thus, rely on different information standards. Military communication systems are produced by a variety of competitive companies that have an interest in keeping their codes secret and, furthermore, have to meet high standards of security. As a result, software codes and communication protocols have become proprietary. Thus, achieving interoperability is a matter of time and effort. This explains the often-long time frames (up to 20 years) in which states envisage their NCW projects to achieve full operational capability.[4]

The value chain goes on to assume that, through the provision of a physical networked information infrastructure, a steady flow of information is provided that leads to an increased shared situational awareness of the units in the theatre and at the command levels. Shared situational awareness is the contextual integration of all relevant operation information by two or more actors. Shared understanding includes the commander's intent and, thus, an extrapolation of future actions. Better situational awareness and understanding, in turn, translates into a war-fighting advantage as units are enabled to better and more quickly synchronise their actions. Self-synchronisation is a central concept of NCW, as self-synchronising units bypass the stove-piped information channels of traditional military hierarchies and, so the promoters of NCW assume, can coordinate their actions better and more effectively.

Better situational awareness, furthermore, leads to more precise military engagement, which means that less troops or materiel needs to be employed to achieve a certain operational aim. Here, NCW is envisaged as an enabler of another transformational concept – Effects-Based Operations (EBO) (Cebrowski 2003b). Under the conditions of EBO, the massing of forces is replaced by the massing of effects (Smith 2002). At the end of the NCW value chain stands increased combat effectiveness.

2.3.2 Domains of Warfare and the NCO Conceptual Framework

Others aspects of NCW theory include its particular theoretical application to war and how war is theorised by the authors. While war and combat undoubtedly are complex situations, the promoters of NCW decompose combat situations to demonstrate the applicability and value of NCW. For this purpose, they apply a domain model of warfare (Alberts et al. 2001). The domain model divides combat along functional categories into the physical domain (in which manoeuvres take place), the cognitive domain (in which

4 For example, the British Ministry of Defence envisaged that its Network-Enabled Capability will have achieved a mature state only by 2025 (Ebbutt 2006).

the perceptions, understanding and beliefs of the soldiers lie) and the information domain (in which information is created and communicated). The domain model, as such, was first issued by J.T.C. Fuller, a British Major General, as early as 1926 (Fuller 1926).

It was John Garstka who adapted NCW to the domain model, arguing that NCW would considerably strengthen the soldier within the information domain, for example, by providing a common operational picture (Garstka 2000; Office of Force Transformation 2003). In the physical domain the application of NCW would lead to networked sensors and shooters. In the cognitive domain the provision of a common operational picture would lead to a shared situational awareness among all individuals involved. This in turn would lead to a shared knowledge of the commander's intent, which again would enable the units to self-synchronise. The three-domain model was extended in 2003 by a fourth domain – the social –, which theorized about the interaction of individuals, for example, when they exchange information or come to joint decisions. The social domain was an innovation of the NCO Conceptual Framework (NCO CF), which was drafted by the Office of Force Transformation (OFT), the Command and Control Research Program (CCRP) and other entities of the U.S. Department of Defense (DoD) (with the collaboration of the *RAND Corporation*)[5] in order to develop the means to validate the hypotheses of NCW that were presented in the CCRP publications and in a report to Congress in previous years (Garstka/Alberts 2004). In NCO CF, the domain model and the NCW value chain were eventually merged. NCO CF, furthermore, introduced a new 'value' in the value chain – the *agility* of military units. Moreover, it developed a specific set of metrics with which to test the assumptions that NCW made. Finally, NCO CF presented the results of the various case studies that have been conducted on NCW

2.3.3 The Roots of Network-Centric Warfare

The term NCW might only have appeared for the first time in 1998, but the development of NCW was preceded by a number of events and doctrinal developments that took place well before or during the 1990s. Any account of NCW that omits its origins would ultimately appear a-historic and presumably equip the reader with a false belief about the NCW diffusion process.

To be sure, NCW was not invented by Cebrowski, Garstka, Stein or Alberts in an academic ivory tower. Rather, it can be traced back to a number of intertwined technological, conceptual, intellectual and strategic roots: Technologically, the information revolution, the advent of stealth, precision and speed; doctrinally, the emergence of cooperative and joint war-fighting concepts; intellectually, the debate concerning a revolution of military affairs; and strategically, the changed mission spectrum of the U.S. Armed Forces after the end of the Cold War. The following sub-sections will briefly touch upon these four issues.

5 The RAND Corporation (RAND stands for *R*esearch *AN*d *D*evelopment) is an influential U.S. think tank originally founded after WWII as a U.S. Air Force project (Abella 2008; Kaplan 1983).

2.3.3.1 Technology

As stated above, NCW is not a technology, it is rather a concept. Nevertheless, technology does play an important role: In the NCW value chain, technology is coined the "entry fee" (Alberts et al. 1999: 187). NCW built on the existence of sensors – i.e. electro-optical sensors to survey a certain territory, GPS sensors, fuel-level sensors in vehicles, etc. –, communication technology, IT and precision weapons; but – if you think of drone reconnaissance – it also benefited from advances in the area of aviation. As scholars correctly point out, these technologies are the results of civilian and military technological developments of the 1960s, 1970s and 1980s (Owens 2001; Sloan 2008; Vickers/Martinage 2004). Furthermore, they were already used before NCW was drafted: Satellites, for example, were used for military reconnaissance and communication purposes already during the Vietnam War by the U.S. military. So was laser-guided ammunition (Tobergte 2006). Yet, the NCW concept put these existing, or emerging, technologies into an unprecedented context by linking them all into a single network.

Presumably the most significant development for NCW were the advances in the areas of C4I – Command, Control, Communications, Computing and Intelligence – which allowed for the faster processing of information and the faster exchange of information, thus making command and control, a decisive military function, more effective. The improvement in C4I was the result of the rapid developments in computer technology that had been labelled the "information revolution" (Nye/Owens 1996; O'Hanlon 2009: 174). The performance of computer chips and, as a result, processing capacity, increased by nearly 2.8 million percent from 108 kHz in 1974 to over 3 GHz in 2007.[6] This allowed for an ever-faster processing of large amounts of data, for converting, compressing and harmonising information. As under the condition of network-centric warfare "information is a key resource in military operations" (Collmer 2007), the increased capacity to transmit and process data has had tremendous effects on the command and control of operations.

It is important to note that the potential impact of information technology on the conduct of war has two dimensions. Whereas increased capacity to process information allows for a better – and in the case of NCW theory, different – conduct of traditional military functions (information gathering, command and control and engagement) the information domain itself can also become a focus of warfare (information warfare and cyber warfare). Both topics are undoubtedly interrelated, but they should not be confused. To a large extent, NCW theory addresses the first understanding of the impact of IT on warfare.

Williamson Murray and Robert Scales (2003: 57) pointedly concluded on the impact of information technology on U.S. military capabilities that: "By the 1990s the computer revolution had reached warp speed and was having an enormous impact in communications (synchronous and asynchronous), data collection, information procuring, intelligence gathering, and speed of calculation. The Global Positioning System, which had been of limited use in 1991 in helping Coalition ground forces find their way around Iraq's deserts, was an integral part of a number of weapon systems by 2003."

6 See Intel Company, http://www.intel.com/technology/timeline.pdf (accessed on 01 August 2012).

2.3.3.2 Cooperative and Joint Doctrines and Concepts

Apart from leaps in technology, new operational concepts that stressed the co-operative networking of systems and joint-force integration (in contrast to a mere collaboration of platforms, or sea, land and air forces) evolved. Since the mid-1970s, the U.S. Navy has promoted research on co-operative targeting by which one ship could engage a target by using radar data provided by another source that was then transmitted via a data link. In 1987, the CEC programme was established (Cooperative Engagement Capability), resulting in concrete acquisition projects in 1992. In various publications, CEC is named as the doctrinal forerunner of NCW (Scott 2003; Vickers/Martinage 2004). This close conceptual connection is furthermore established by the fact that the first mentioning of the term Network-Centric Warfare occurred in the context of the U.S. Navy's digitisation programme Copernicus, of which CEC formed one substantive part.

Complementary to the Navy's efforts, the AirLand Battle Doctrine emphasised the close coordination of air and land operations and deep strikes to prevent the enemy (the Warsaw Pact) from reaching the frontline. The AirLand Battle Doctrine was conceptualised in 1976 and formally proclaimed in 1982 by the U.S. Army (Lock-Pullan 2005). On the tactical level, the Army long-since relied on Close Air Support (CAS) and Joint Tactical Fires (JTF), and now NCW was a means to improve these concepts. Conceptually, NCW was furthermore influenced by the idea of a 'system of systems'. In the mid-1990s, Admiral William Owens made initial assumptions concerning the integration of reconnaissance, C2 and precision on the battlefield regardless of service boundaries (Owens 1995, 1996, 2001).

2.3.3.3 The RMA Debate

The early 1990s saw a debate on the influence of the information revolution on societies and, ultimately, on the conduct of war (Owens 2001). This debate was preceded by the writings of futurists of the 1980s who anticipated a profound change in the economy and society due to the invention of the computer and proclaimed the advent of the "information age" (Naisbitt 1982; Naisbitt/Aburdene 1985; Toffler 1980). The writings of Alvin and Heidi Toffler proved decisive for the implications that originally civilian information-age technologies and business processes might have for the conduct of war (Toffler/Toffler 1993). The Tofflers drew a picture of the Information Age as comprising of the growing relative importance of knowledge and information as a production factor, the intangible value of companies able to produce certain kinds of knowledge or information and to generate ideas, the de-massification of production, labour specialisation, constant innovation, flexibility in production processes and scale, a networked organisational structure and accelerated business processes (Toffler/Toffler 1993: ch.8). They went on to show how parallel developments occurred within the U.S. armed forces and their way of making war.

When new technologies, such as the Global Positioning System (GPS), Tomahawk cruise missiles and laser-guided bombs were used, and joint war-fighting concepts, such as AirLand Battle, were applied by the U.S. military during Operation Desert Storm in 1991, some military thinkers acknowledged this as a caesura where the conduct of conventional military operations by the U.S. Armed Forces moved from industrial-age war to information-age war (Krepinevich 1994; Maynard 1993; Perry 1991; Ritcheson 1996; Sullivan/Dubik 1993). "Something occurred in the night skies and desert sands of the

Middle East in 1991 that the world had not seen for three hundred years – the arrival of a new form of warfare (…)" (Toffler/Toffler 1993: 73).

This new form of war, it was argued, was no longer characterized by an emphasis on firepower and the massing of heavily-mechanised troops along a pre-defined frontline, but by small and agile forces, moving speedily over large distances, penetrating deep into the enemy's territory, wirelessly exchanging mission-critical information while already being on the move and ready to take action with unprecedented precision. Special emphasis was put on the role of C3I technology, resulting in increased situational awareness and on precision-guided weapons that acted as force multipliers and considerably increased the effectiveness of the forces (Mahnken 2000; Perry 1991). Operation Desert Storm became the scenario for future war planning and guided U.S. military reform. Some thus referred to the campaign as "the mother of transformation".

Not everybody, however, was convinced that there was indeed a new way of conducting war. A debate occurred in the mid-1990s about whether or not the first Gulf War demonstrated a major leap in technology and operational concepts and scholars debated if the 1991 victory was simply "misunderstood" (Biddle 1996, 1997; Keaney 1997; Mahnken/Watts 1997; Press 1997).

Others, while not outright denying the impact of technology on victory in Operation Desert Storm, were nevertheless critical of the lessons drawn thereafter, especially the increased focus on technology in U.S. military change (Murray 2003). Others even argued that the overwhelming success of Operation Desert Storm resulted in a "residual overconfidence" that "validated existing doctrines" (Owens 2001: 19) and might have "inhibited more innovative thinking about the RMA in the American military in general" (Rosen 2010: 469). The change of operational art and tactics believed to be seen in Operation Desert Storm, nevertheless fuelled the more general debate concerning the Revolution in Military Affairs that occurred in the U.S. in the late 1980s (Murray 1997).

An early and influential contributor to the RMA debate was Andrew Marshall, who – as the director of the Office of Net Assessment in the U.S. Department of Defense – not only coined the phrase Revolution in Military Affairs and spread the idea that there indeed was a military revolution going on, but who also sponsored a number of studies on historic military innovations and lessons drawn from them for the current process of military change (Krepinevich 2002 (1992); Murray/Millett 1996).[7] In 1994, Andrew Krepinevich, then President of the Center for Strategic and Budgetary Assessments, published an influential article in the *National Interest* in which he demanded that the current military advantage America had in terms of technology and military systems must be actively defended through further innovation and adaptation against competitors (Krepinevich 1994). Other early contributions to the RMA debate focused on the potential future changes in warfare. "Cyberwar is Coming" was an often-cited article authored by RAND analysts (Arquilla/Ronfeldt 1997 (1993)) that opened a debate on how warfare might look like in the future and how the U.S. armed forces would have to adapt in order to maintain their military superiority (Nye/Owens 1996). Of course, the assumptions made by RMA proponents were not uncontested. Eliot A. Cohen, for ex-

7 The role of Andrew Marshall is discussed in-depth in the article "The Impact of the Office of Net Assessment on the American Military in the Matter of the Revolution in Military Affairs" by Stephen P. Rosen (2010). For a brief overview, see Thomas Mahnken (2008: 176f.). The collaboration between the Office of Net Assessment and the military historians Allan Millet and Williamson Murray is also mentioned in Lynn (2001).

ample criticised the RMA debate as not being placed within a geopolitical context, focussing too much on technology while, at the same time, neglecting organisation, doctrine and manpower (Cohen 2004).

In some respects, the Kosovo fuelled the RMA debate. Politically, it might be discussed whether Milosevic finally withdrew Serbian forces because of the threat to employ American and British ground forces (Blair 2010), because of the success of NATO's air campaign or due to the decline of Russia's political support (Arkin 2001; Mahnken 2008: 187f.). From an operational point of view, the Kosovo campaign seemed to verify some basic tenets of the RMA – the usefulness of surgical, long-range precision strikes and enhanced C4ISR (Vickers 2001). Yet, William Owens pointed out that the technological potential the U.S. military had in Kosovo was not accompanied by a suitable strategy and force structure and thus inefficiently used (Owens 2001). The debate on an RMA and the vision of an approaching information age served as the intellectual frame, not only for the U.S. military transformation project that took off in 2001, but also, more specifically, for the formulation of NCW-theory. In their initial writing on NCW, Arthur Cebrowski and John Garstka made reference to the RMA (Cebrowski/Garstka 1998).

2.3.3.4 Strategic Changes after the Cold War

In the 1990s, it became apparent that the sources of new security threats were uncertain. This was in contrast to the situation before the 1990s when, in the bipolar international system, the military strength of the Warsaw Pact was more or less anticipated and formed the basis on which to build NATO defence planning. The new strategic uncertainty, the rising number of regional conflicts in Africa and on the European continent led to the questioning of the former threat-based approach to military planning. Instead, military officials started to think about what capabilities the military should have in order to be prepared for the most likely future military tasks and for smaller missions of higher frequency. In a testimony before the Senate Armed Services Committee, then Secretary of Defense Donald Rumsfeld explained that "uncertainty about the future strategic environment" was best faced by "capabilities-based" planning, "to make certain we develop forces prepared for the longer-term threats that are less easily understood" (Rumsfeld 2001a).

A decisive document that merged the strategic changes after the end of the Cold War on the one hand with the ongoing debate concerning the impact of technology on the other was the *Joint Vision 2010* (JV 2010, Shalikashvili 1996). Published by the U.S. Joint Chiefs of Staff, the document was a preparation for the Quadrennial Defense Review (QDR) in 1997. The QDR is a Congressionally-mandated assessment of all aspects of U.S. defence that is released every four years. It aims at reviewing existing military doctrines, organisation and capabilities that fit the strategic environment and to adjust them when necessary.

The end of bipolarity with the Warsaw Pact, it was argued in the JV 2010, could bring about new kinds of adversaries with different capabilities and conflicts with different levels of escalation. Therefore, the JV 2010 proposed a new strategic imperative for the U.S. Armed Forces doctrine: *Combat superiority* was to replace the previous doctrine of (balanced) mutually assured destruction of Soviet forces. The new concept of combat superiority was based on *full-spectrum dominance* in the theatre and it drew on technological superiority, information superiority – which would provide dominant battlefield awareness – and the jointness of forces. The vision was to replace the existing Cold War-style military logic of "massed forces, and sequential operations" (Shalikashvili 1996: 17) with an approach that aimed at using C3I technology and precision-guided weapons to achieve desired effects. Ultimately, the JV 2010 was built on the idea that information technology would allow for a new quality of joint operations.[8]

To conclude, NCW formulation was shaped by four major developments. Technological advances from the 1970s onwards, especially information and communication technology, enabled developments in the realm of military command and control. Together with innovations in sensor technology, aviation and precision ammunition, these advances provided the basis upon which NCW, as a new organisational concept of war, was formulated. Furthermore, the trend to act jointly in combat resulted in the need for innovative tactical and operational concepts for inter-service communication. The RMA debate, moreover, stimulated the formulation of original concepts, as it promoted a fresh way of approaching military challenges by looking for innovative trends in the broader society. Finally, the end of the Cold War and the new uncertainty in international security resulted in the necessity of adapting the way American forces fight. While the formulation of NCW can be traced back to these four developments, the ultimate success of the concept in terms of popularity within military circles was almost intrinsically tied to the U.S. military transformation since 2001. Thus, the next section describes how NCW took off and quickly became a pillar of the military transformation during the term of Secretary of Defense Donald Rumsfeld (2001–2006).

2.4 The U.S. Military Transformation Project

2.4.1 The Transformation Rhetoric

In the current debate, the term 'military transformation' can refer to three not entirely separated phenomena. In its most general meaning, transformation was used by scholars and security experts to describe profound military change in the very nature of war, in the composition of the military, in military strategy or civil-military relations (van Creveld 1991). Others – while also employing a rather descriptive notion of military transformation – limited their observations to technological innovations and how they changed the conduct of war. This understanding of military transformation had its heyday in the late 1990s in conjunction with the debate about a revolution in military affairs and the information revolution (Sloan 2008: 7f.). Still others employed a less descriptive and more utilitarian view on military transformation defining it, for example, as "the act of creating and harnessing a revolution in military affairs" (Binnendijk 2002: xvii). Finally, during the tenure of Secretary of Defense Rumsfeld, the term transformation (figuratively with a capital T, see Osinga 2010) more specifically referred to the defence-wide "high-technology transformation" (Tangredi 2002: 8) project the Pentagon, the U.S. strategic command and the U.S. individual services pursued. This latter notion of U.S. military transformation was the least descriptive, as it had real policy implications and a significant impact on U.S. military procurement projects, doctrine and operations. Transformation thus was a political dictum, a policy goal.

8 A number of joint military documents further developed the JV 2010 concepts, i.e. the Concept for Future Joint Operations (U.S. Joint Chiefs of Staff 1997) and the JV 2020 (Shelton 2000).

When U.S. military transformation took off in 2001, proponents and supporters pointed at a number of specific features that would distinguish the transformation project from a mere modernisation of forces: Transformation, they remarked, was not solely the introduction of new technologies, but also comprised of major changes in doctrine and force structure and the development of new war-fighting concepts (Krepinevich 2002). Furthermore, transformation was defined as a continuous process of constant adaptation to changing circumstances rather than having a defined end-state (U.S. Department of Defense 2003).

2.4.2 U.S. Military Transformation

The U.S. military transformation was the attempt by U.S. defence policy-makers to bring size, structure, doctrine and equipment of the U.S. armed forces in line with the changing strategic environment after the end of the Cold War. Even though the term 'military transformation' emerged as a buzzword and a tangible policy project only during the Bush Administration (2001–2009) and, more specifically during the tenure of Secretary of Defense Rumsfeld (2001–2006), the processes that ultimately led to the announcement of a U.S. military transformation had already started during the Clinton years (1993–2001) when the need for the U.S. military to adapt to the changed strategic environment after the Cold War was recognised by the administration (Tangredi 2002: 8).

In 1996, the JV 2010 envisaged a new American military that would gain combat superiority through the exploitation of U.S. technological superiority for military purposes. In May 1997, the QDR was released, acknowledging the impact of IT on war and used terms like *RMA* and *transformation* to refer to the upcoming changes in the U.S. military (Cohen 1997: section VII). Ultimately, the "transformation" rhetoric was used as a replacement for references to the "RMA" in the report of the National Defense Panel on "Transforming Defense" of December 1997 (U.S. National Defense Panel 1997). A number of initiatives that followed aimed at harnessing the advantages of the information revolution and putting the JV 2010 into practice. In 1998, for example, Secretary of Defense Cohen tasked the U.S. Atlantic Command (a year later to become the Joint Forces Command) to lead joint concept development and experimentation in the areas of C2 and networking (Cohen 1999).

2.4.3 The New Administration and the Transformation Agenda

In the run-up to the Presidential elections, candidate George W. Bush delivered a widely recognised speech to The Citadel military academy (Bush 1999; O'Hanlon 2002). In this speech, Bush outlined how he would transform the U.S. military into a force of the next century by exploiting information technology With Donald Rumsfeld, President Bush chose a Secretary of Defense that shared his own view on the future shape of the military and embraced the NCW concept.

Rumsfeld was appointed Secretary of Defense in May 2001. One of Rumsfeld's first steps as Secretary of Defense was to convene a group of military and non-military experts to study how a military transformation could best be achieved. The group, also known as the McCarthy Panel, presented its works in April 2001 and thus influenced respective sections of the upcoming QDR that was due in September 2001 (McCarthy et al. 2001), when Rumsfeld presented his vision of a profound military transformation (Rumsfeld 2001b). In particular, the QDR outlined a number of operational goals for the single services to be achieved in the next few years: Protecting critical bases and defeating chemical, biological, radiological and nuclear weapons; projecting and sustaining forces in anti-access environments; denying enemies sanctuary; leveraging information technology; assuring information systems; conducting information operations and enhancing space capabilities (Rumsfeld 2001b: 30). Furthermore, Rumsfeld envisaged a specific way in which these six operational goals could be achieved – through increased jointness, through better exploitation of U.S. intelligence advantages, through the method of concept development and experimentation and, finally, through developing transformational capabilities. The QDR referred to these four mechanisms as the "transformation pillars" (Rumsfeld 2001b: 32). For the sake of completeness, it is important to note here that Donald Rumsfeld's military transformation agenda also addressed a number of other concerns that were non-operational, such as the transformation of business practices within the Pentagon and Homeland Security (Sloan 2008: 46).

Bold in its outlook, Rumsfeld's defence vision received, however, rather negative comments and provoked resistance from the U.S. Army that was already pursuing a modernisation programme of its own. O'Hanlon, for example, argued that the 2001 QDR was the least programmatic one (in terms of force structure and mission spectrum changes) and was rather rhetorical in nature (O'Hanlon 2002: 105). Yet, Rumsfeld, to some degree, managed to put some of the transformation plans into practice. For example, he cancelled major legacy projects such as the Crusader artillery system in 2002 and the Comanche helicopter in 2004 and – as will be discussed below – introduced NCW into the services. Rumsfeld's core belief about technology was also criticised. His transformation vision (and later his approach to fighting the war in Iraq), to a large extent, relied on the exploitation of new technologies at the expense of manpower. In the beginning, this view was most strongly opposed by the manpower-strong U.S. Army that feared becoming marginalised while the Air Force and the Navy could hope to benefit from a technology-driven plan to modernise the armed forces. An observer of the Pentagon business remarked (Scarborough 2004: 29): "The generals were beginning to learn that, while 'everybody likes boots on the ground', as Rumsfeld put it, the new defense secretary was looking for more creative war plans. He wanted a fast flow of battlefield intelligence, precision air strikes, and light, lethal units on the ground."

Later, Rumsfeld's conviction came under attack again, as it proved difficult during the Iraq War (Weinraub/Shanker 2003). Finally, and somewhat detached from the substantive issues at stake, Rumsfeld's particular personality and his style of management might also have raised opposition from the military and from the wider defence system (see Scarborough 2004: 112ff.): The events that took place in the aftermath of the terror attacks of 11 September 2001 diminished the strong opposition against the technology-focused military transformation for a while. In the reading of Rumsfeld and transformation proponents, the Afghanistan War confirmed the RMA hypothesis as the combination of unmanned aerial vehicles, special operation forces (on horseback), and precision ammunition seemingly brought about a sweeping victory (Rumsfeld 2002b). For some – and in contrast to some critical reviews that also appeared – this view was further reinforced by the U.S. victory during the invasion of Iraq in 2003 that seemed to substantiate the assumption "that the integration of air power, special forces, and smaller, more mobile ground forces could enable the U.S. military to be more effective with less mass" (Sloan 2008: 139). As a result of this 'validation' of the RMA idea in combat, the transformation project gathered steam. Rumsfeld cancelled some legacy systems and increased funding of transformational systems, such as unmanned aerial vehicles and precision-guided missiles (Sloan 2008: 140).

In 2003 a pamphlet with the forceful title of "Military Transformation: A Strategic Approach" was issued (Office of Force Transformation 2003). The paper restated the operational goals and transformational pillars from the QDR. Furthermore, it discussed the significance (and success) of the transformation against the background of the combat operations in Afghanistan and Iraq. Finally, it advertised a new way of thinking about military change and propagated an open-ended process of continuous adaptation (Office of Force Transformation 2003: 8): "First and foremost, transformation is a continuing process. It does not have an end point. Transformation anticipates and creates the future and deals with the co-evolution of concepts, processes, organizations, and technology (…). The overall objective of these changes is to sustain U.S. competitive advantage in warfare."

This new thinking clearly put into question many long-established methods in the management of defence. Rumsfeld thus had to make sure to make the services comply with the transformation agenda. To this end, in late autumn 2001 he created the Office of Force Transformation (OFT) – a "think and do tank" (Blaker 2007: 5) – to centrally steer and oversee military transformation in the individual services. Tasked with overseeing transformation, the OFT was perhaps the most critical transmission belt for the introduction of NCW in the U.S. Armed Forces.

2.4.4 The Office of Force Transformation

The OFT was an organizational subunit of the Office of the Secretary of Defense and thus hierarchically located at the top of the Pentagon. It comprised of 20 personnel. The director of the OFT became no other than Arthur Cebrowski, the co-author of the original NCW-article in 1998. Cebrowski was convinced that in order to push the transformation vision, the existing mind-set had to change. During his tenure he thus aimed at accomplishing the following goals (quoted in Roxborough 2002: 71):

- "Make force transformation a pivotal element of national defense strategy and DoD corporate strategy, effectively supporting the four strategic pillars of national military strategy.
- Change the force and its culture from the bottom up through the use of experimentation, transformational articles (operational prototyping), and the creation and sharing of new knowledge and experiences.
- Implement Network Centric Operations (NCO) as the theory of war for the Information Age and the organizing principle for national military planning and joint concepts, capabilities, and systems.
- Get the decision rules and metrics right and cause them to be applied enterprise-wide.
- Discover, create, or cause to be created new military capabilities to broaden the capabilities base and mitigate risk."

To achieve this, the OFT had a wide-ranging mandate to oversee the transformation efforts of the individual services: "The Director, Office of Force Transformation (OFT), will monitor and evaluate implementation of the Department's transformation strategy, advise the Secretary, and manage the transformation roadmap process" (U.S. Department of Defense 2003: 12). The OFT developed a formal policy process – the TPG-process – with which to monitor and steer the transformation in the services. The acronym TPG refers to the Transformation Planning Guidance – a document published in April 2003 that elaborated on the scope of transformation ("Transforming How We Fight", "Transforming How We Do Business", "Transforming How We Work With Others") – that requested the services to produce roadmaps (U.S. Department of Defense 2003). Additionally, with the TPG, the OFT tried to establish a common understanding among all key defence institutions about transformation and its basic concepts, such as NCW.

2.4.5 The Transformation Planning Guidance Process

The TPG listed a number of short- and mid-term targets for the services and strategic commands to achieve. The U.S. Joint Forces Command (USJFCOM), for example, was tasked with developing joint doctrines for war fighting, concept development and experimentation. As a result, the USJFCOM issued a Joint Operations Concept (JOpsC) in November 2003. This in turn influenced the reformulation of subordinate documents such as the Joint Operating Concepts (JOC), Joint Functional Concepts (JFC) and Joint Integrating Concepts (JIC).[9]

The individual services were tasked with developing so-called transformation roadmaps. These roadmaps should – on the basis of the transformation definition issued by the DoD – identify what military capabilities could be fielded, when and how; what critical links to the other services and agencies existed (for example, in the area of strategic lift) and what changes in organisational structure, operating concepts, doctrine and training would be necessary to achieve more agile and more effective forces.

The OFT envisaged a cyclic approach of monitoring: The single services were asked to issue their roadmaps annually by November. The OFT then evaluated the efforts in the Strategic Transformation Appraisal. Although it was feared that the TPG-process would duplicate existing monitoring processes, all individual services issued the requested roadmaps. Comparing the actual service roadmaps with the TPG, Mark Czelusta found that the Army stuck closest to the guidelines (Czelusta 2008). In contrast, the Air Force and Navy not only had developed their own definitions of the transformation process, but also had partially ignored tasks issued in the TPG. According to Czelusta, the reason lay in the legacy of the past. During the Clinton administration, the services had already started their own modernisation programmes and thus had already chosen a specific path towards transformation. The Army, for example, had an organisational focus. It aimed at breaking up heavy divisions, replacing them with smaller, digitised, more agile brigades. The Air Force, in contrast, employed a rather technological view, stressing the need of developing tactical data links.

The USJFCOM even added more differentiation to the process. In its Joint Transformation Roadmap issued in January 2004, it did not develop a strategy for combining or integrating the transformation efforts of the Army, Navy or Air Force. Instead, it

---

9 These documents are developed in the J7 Joint Experimentation, Transformation, and Concepts Division (JETCD) of the USJFCOM. Examples of JOC are "Stability Operation JOC v.1" or the "Major Combat Operations JOC v. 1" both from September 2004 (see http://www.dtic.mil/jointivison/joc.htm accessed 04.02.2009). Joint Functional Concepts are more specific and comprise, among others, of "Focused Logistic JFC" of December 2003 or the "Net-Centric Environment JFC" of April 2005 (http://www.dtic.mil/futurejointwarfare/jfc.htm accessed 4 February 2009).

added new concepts to the already muddled discussion about transformation (U.S. Joint Forces Command 2004). Still, despite the initial impediments, the transformation finally took off and the comprehensive doctrinal and technological military transformation became a top priority in each of the single services, the military commands, and ultimately, in NATO and among NATO allies. A variety of reasons account for this: the initial successes in the combat operations in Afghanistan, and later in Iraq, seemed to validate the techno-centric transformation approach, the increases in defence spending and probably the attractiveness of new phrases, such as *network-centric* or *effects-based*, that pervaded uncounted power point presentations during the early years of the new century.

2.4.6 The Cooling Down of the Transformation

The TPG process did not carry on for long – in August 2005, the DoD stopped the TPG process. In the same year, Arthur Cebrowski had stepped back from his position as the director of the OFT due to health reasons; and in Autumn 2006, the Office of Force Transformation was dissolved as an independent subunit and had been incorporated into other DoD units. In November 2006, Rumsfeld resigned as Secretary of Defense and was replaced by Robert Gates. With Rumsfeld resigning from office at the end of 2006 and the resignation of Arthur Cebrowski from the post of director of the OFT due to his illness in 2005, the U.S. military transformation process had lost its two key personnel drivers.

Since the post-invasion phase in the Iraq war, Rumsfeld's vision of a technology-driven armed forces transformation had come under heavy pressure.[10] Surely, critical comments on the transformation project had been issued since its very inception. Yet the Secretary had lost the support of the services. One anecdote is telling in how Rumsfeld had lost touch with the troops: In early December 2004, Rumsfeld visited National Guard troops in Kuwait. Soldiers complained about the lack of appropriate equipment, pointing at the fact that they were forced to dig through landfills in order to find scrap metal to reinforce their vehicles. Apparently, the networked, mobile and light infantry units had suffered from losses due to insufficient armour against roadside bombs. In frustration, Rumsfeld responded to the soldiers' complaints: "You go to war with the Army you have."[11] His remark led to a major public debate in the U.S.

Ultimately, Rumsfeld's approach to the war (small force) and his overall approach to transformation (replacing manpower with technology), sometimes called the 'Rumsfeld Doctrine', came under pressure. A number of retired Generals called for Rumsfeld's dismissal in 2006, blaming him for his "absolute failures in managing the war against Saddam in Iraq" (General Swannack quoted in Cloud/Schmitt 2006). By 2005, the U.S. Army, carrying the major weight of the occupation of Iraq, started to emphasise counterinsurgency over net-centricity (see Field Manual 3-24 [Counterinsurgency]) and irregular warfare was accounted for in the 2005 QDR [Graham 2005]). In December 2006, the Iraq Study Group – a group of experts including former Secretary of State Baker – presented its report on the situation in Iraq and ways to change the constantly

10 See, for example, the New York Times' editorial "A Science Fiction Army", 31 March 2005.
11 See various newspaper articles from 10 December 2004 (i.e. The Washington Post).

worsening security situation (Iraq Study Group 2006). The Group argued for a temporary, but significant *increase* in force size, therefore rendering Rumsfeld's approach obsolete. This recommendation was put into practice with the troop surge in 2007.

2.4.7 NCW and the U.S. Military Transformation

Usually, there are three broad major changes associated with the current wave of military transformation in the U.S. and, subsequently, in the military organisation of a number of other states: effects-based operations, expeditionary forces, and, network-centric warfare (Terriff et al. 2010). With the father of NCW, Arthur Cebrowski, becoming director of the OFT and thus being personally responsible for the transformation process, it seemed almost inevitable that NCW would have some role to play in the U.S. military transformation. As a matter of fact, NCW became a centrepiece of U.S. armed forces' transformation and was even officially recognised as nothing less "than the embodiment of an Information Age transformation of the DoD" (U.S. Department of Defense 2001). The importance of C3I, and especially of achieving common situational awareness, had already been articulated in the JV 2010. During the 1990s, the individual services had started to pursue their visions of how digitisation would benefit their military tasks. Yet, apart from the JV 2010 (and its follow-up JV 2020 [Shelton 2000]), a defence-wide concept for the joint exploitation of C3I across service boundaries was lacking. NCW filled that gap.

With a view to the upcoming QDR, Congress tasked the Pentagon in 2000 to look into the issue of NCW (see Section 934 of the Defense Authorization Act for Fiscal Year 2001). The QDR 2001 made some explicit reference to NCW by acknowledging the Department's current transition to network-centric warfare, although it might be surprising that NCW was not very prominent in the paper. There were only some implicit references to the NCW theory (Rumsfeld 2001b): "[N]ew information and communications technologies hold promise for networking highly distributed joint and combined forces and for ensuring that such forces have better situational awareness; [and] the capability provided by this network and its applications will enable rapid response forces to plan and execute faster than the enemy and to seize tactical opportunities."

This was presumably due to the fact that large parts of the QDR had been worked on during the tenure of Rumsfeld's predecessor, William S. Cohen. Yet, this is not to say that NCW had not sparked considerable interest in the defence area before. In March 2001, for example, the USJFCOM issued a report on NCW and Joint Experimentation. Furthermore, the DoD had presented to Congress a report on its plan to make NCW the cornerstone of the armed forces transformation in July 2001 (U.S. Department of Defense 2001). The NCW report was the basic document by the DoD outlining the proposed course of action with regards to introducing NCW into the U.S. military. Yet when Arthur Cebrowski arrived at the Pentagon, the relevance of NCW multiplied. In fact, achieving network-centricity soon became one operational aim of the transformation effort; in the foreword to the transformation pamphlet, Cebrowski stated that the U.S. is "on course to transform [its] military into an agile, network-centric, knowledge-based force capable of conducting effective joint and combined military operations against all potential future adversaries" (Office of Force Transformation 2003).

2.4.8 On the Prominence of NCW

NCW was a cornerstone of the U.S. military transformation project. Yet, it was argued above that NCW was rooted in a number of technological, strategic, intellectual and doctrinal developments that started way before 2001. One cannot help asking why earlier and similar conceptions, such as the system-of-systems approach by William Owens, vanished from the debate without substantial impact on policy and why NCW was so prominent that one could almost speak of an *NCW-hype* between 2001 and 2005. Apparently with a new incoming administration, NCW was the right concept at the right time, sponsored by the right influential people.

It was the tandem of Donald Rumsfeld, the "Transformation Czar" (quoted in Binnendijk 2002: xix), and Arthur Cebrowski, father of NCW and also "the prime architect of U.S. military transformation" (Blaker 2007: x), that is fascinating in this regard. Rumsfeld – although he largely earned rather critical appraisals – is widely considered as being one of the most powerful Secretaries of Defense (Blaker 2007: 23) in history. He was a believer in the revolution in military affairs and was convinced that information dominance and precision strikes would ultimately lead to a decreasing need for 'boots on the ground'. When Rumsfeld arrived in the Pentagon, he soon started working on the QDR, and assigned the task of oversight to Andrew Marshall, the director of the Office of Net Assessment and most senior proponent of the RMA idea (Sloan 2008: 139).

Some months later, Rumsfeld offered the job as the director of the newly established Office of Force Transformation to Cebrowski. As mentioned above, Cebrowski was a retired Vice Admiral and former President of the Naval War College. His background as a Director of Communications in the Joint Chiefs of Staff, his analytic style and his eagerness to change the culture of inertia in the services and the Pentagon destined him to become "Rumsfeld's handpicked aide for transforming the armed forces to fit Rumsfeld's mould" (Scarborough 2004: 44). A less flattering picture of Arthur Cebrowski is drawn by P.W. Singer, who quotes from an article by Clay Risen, in which Cebrowski is described as an "obsessive Power Pointer" having a "nervous energy and a maverick streak that made him prone to trendy theories and sky-high philosophizing" (Risen 2006; Singer 2009: 179f.). Yet, neither Singer nor Risen question the role of Cebrowski in promoting NCW.

Apart from Cebrowski, two other NCW pioneers – John Garstka and David Alberts – also received high positions in the Pentagon. Garstka became the Chief Technology Officer for the Joint Chiefs of Staff and Alberts became the Director of Research and Strategic Planning for the Assistant Secretary of Defense for Command, Control, Communications and Intelligence (Book 2002). Garstka became well known as the project leader of the various NCO case studies that sought to validate NCW theory.[12] Alberts acted as the director of the Command and Control Research Program (CCRP),[13] becoming the leading figure that spread the idea of NCW beyond the sphere of the U.S. military: The CCRP published all of the original volumes on NCW theory and made them available through the Internet. Furthermore, the CCRP hosted high-profile annual international conferences on C2 issues.

12 A comprehensive overview about the case studies, including a short summary about the respective findings, is provided by (Garstka/Alberts 2004: Ch. 7).

13 See http://www.dodccrp.org/ (accessed 2010-09-30).

In addition to the high-level support, a general openness towards information technology developed in the U.S. military. Williamson Murray and Robert Scales noted that, while the Vietnam War generation of officers retired in the 1990s, the new generation of officers was "enthusiastic about the possibilities of technological change for its own sake" (Murray/Scales 2003: 57). Apart from the strong support of powerful Pentagon officials, NCW also became an initial success (in terms of widespread awareness of the concept) and a beacon project for the transformation as it provided a tangible concept to the otherwise rather fuzzy idea of a military transformation. NCW was thus simply better suited to activate support and attention than transformation, as the following quote shows (Cebrowski quoted in Blaker 2007: 180): "So, transformation became the banner under which those opposing rapid change could get away with it. They were able to stretch the term to justify everything they were doing. 'Transformation' had become much less useful as a means of bringing about the rates of changes we believed were necessary. That was why we in the Force Transformation Office began to push the concept of network centric operations. We needed to supplement 'transformation' with a more narrowly focused concept that would maintain the pressure for more rapid change. That meant it had to be more than just a new slogan."

Frederick Kagan (2006: 265) argued in a similar, yet more critical way: "Network-Centric Warfare was thus neither revolutionary nor more accurate than previous concepts (…) had been, and it would probably have had little more effect and staying power than those concepts but for an accident of politics". The Office of Force Transformation was determined to implement NCW into the individual services. It used the TPG process that is described above to achieve a common understanding of the concept and further stimulated individual services' activities to become net-centric.

2.5 Implementing NCW

The individual services had started to look into the issues of digitisation and exploiting C3I technologies long before Rumsfeld's military transformation and Cebrowski's advocacy for NCW. However, a joint approach was missing: The services pursued their own views on NCW. The Navy developed its FORCENet approach, the Army referred to Battle Command (esp. Battle Command on the Move) and the Air Force focused on networking its formerly disparate air and space operations centres (Czelusta 2008: 31).

The OFT tried to harmonise existing NCW programmes, and the TPG requested the services to "explicitly identify initiatives undertaken to improve interoperability in the following areas: deployment of a secure, robust and wide-band network; adoption of 'post before process' intelligence and information concepts; deployment of dynamic, distributed, collaborative capabilities; achievement of data-level interoperability; and deployment of 'net-ready' nodes of sensors, platforms, weapons and forces" (U.S. Department of Defense 2003: 29). Furthermore, it proposed the development of future joint concepts for achieving NCW benefits, such as shared situational awareness, self-coordination and dispersed and de-massed forces (U.S. Department of Defense 2003: 31f.).

The Department of Defense, as the top of the military hierarchy, not only attempted to organise and harmonise services' NCW efforts. The DoD's Horizontal Fusion Program of 2003 aimed at creating the overall network structure in which the subordinated military entities could then link their own systems (Williams 2003). The Global Information Grid (GIG) and the Joint Tactical Radio System and Satellite Communications

(SATCOM) have been the most important projects within the Horizontal Fusion Program. By 2008 however, the program was in doubt as funding dropped constantly (Strassmann 2008) and attention changed towards the field of cyber security and defence (Kenyon 2008). The attention for cyber security had increased and the problem received major attention in the QDR 2010.

Among the individual services, the Army was particularly challenged by the information revolution, contrary to the U.S. Navy and the U.S. Air Force. The latter operated a straightforward number of platforms (vessels, carriers, combat and transport aircrafts and bases). For the Army, however, every infantryman and ground vehicle constituted a network-entity, making the NCW project a much bigger enterprise for the Army than for the Navy and the Air Force. Nevertheless, the network-idea found some high-level support. The U.S. Army White Paper of 2001, for example, envisaged for the Army that "platform designs [would have] an arrangement of system-of-systems technologies [that would] enable decisive manoeuvre, horizontal and vertical, day and night, in all terrain and weather conditions" (U.S. Army 2001: xi). Chiefs of Staff for the Army General Eric K. Shinseki (1999–2003) and his successor General Peter Schoomaker (2003–2007) both focussed on the setting up of a network, especially on improving the Army's C2 assets (Lawlor 2007).

The U.S. Army launched two large programmes in order to spur its digitisation and networking efforts – the Future Combat System and LandWarNet. The Future Combat Systems (FCS) certainly was the beacon project of U.S. Army transformation. Paralleling the aforementioned changes in force structure, FCS aimed at replacing existing heavy vehicles and weapons systems with a "family of new lightweight multipurpose vehicles, sensors and weapons systems" (Kenyon 2007a). Despite cutbacks in 2009, the FCS remained the U.S. Army's largest procurement project. LandWarNet linked the U.S. Army into the Department's Horizontal Fusion Program. It consisted of technological infrastructure (network and applications) and related services (information management and security) (Ackerman 2008).

NCW had a less dramatic impact on the Navy than on the Army. Due to a different functional logic, the Navy had followed a networked approach since decades already and shifted from a platform-centric to a network-centric doctrine (O'Neil 2002: 142; Sloan 2008: 7). The Navy pursued its ForceNET approach, striving to establish a fully networked naval force.[14] Furthermore, it proposed sea-basing concepts (stressing the importance of naval contributions to land operations), emphasising the need for joint C2 interoperability (for the debate on sea-basing, see Till 2006). The Cooperative Engagement Capability (CEC), a high-speed data connection, links naval ships and naval aircraft into a single integrated air-defence network (O'Rourke 2005). The Navy matched these procurement decisions with organisational realignment by the establishment of the Network Warfare Command (NETWARCOM) in Norfolk, Virginia (Lawlor 2006).

Similar to the U.S. Navy, the U.S. Air Force centralised its command and control of operations by establishing the U.S. Air Force Network Operations Command (AFNETOPS) in July 2006 (Lawlor 2006). Furthermore, the U.S. Air Force started to embrace unmanned aerial vehicles such as Global Hawk and Predator. The QDR 2006 estimated that by 2025 nearly 45 percent of the future long-range strike force will be unmanned (Rumsfeld 2006: 46); however, the procurement of manned aircraft remained the top priority. Conceptually, the Air Force emphasised effects-based operations

14 http://forcenet.navy.mil/fn-definition.htm (accessed 2010-09-29).

(EBO) and NCW would be the means by which to achieve this aim. Technically, the Air Force focussed on integrating the existing data collection systems Airborne Warning and Control System and the Joint Surveillance, Target Attack Radar System (AWACS and JSTAR). Another Air Force project was Link 16, a tactical data-link used for automated machine-to-machine exchange of data (radar tracks, platform status, imagery, etc.). Similar to the Navy's CEC venture, the Air Force developed Network-Centric Collaborative Targeting (NCCT), which aimed at integrating ISR data of various kinds to produce a real-time common operational picture that would allow time-sensitive targeting (Mahaffey; Skaar 2005: 65).

2.6 NCW and the Military Operations in Afghanistan and Iraq

The initial combat operations in Afghanistan and Iraq seemed to validate the NCW concept (Osinga 2010: 27f.). Many considered the offensive in Afghanistan in 2001 a success of small and well-equipped troops, a satellite-based communications network, precision-guided weapons and information dominance, especially through the use of unmanned aerial vehicles (Murray/Scales 2003: 58). Donald Rumsfeld concluded in 2002 that the experiences in the Afghan campaign reinforced the need for the kind of military transformation that he had announced (Rumsfeld 2002b: 26). Similarly, the initial phase of Operation Iraqi Freedom seemed to validate the idea that a transforming U.S. military was able to fight a "new way of war" (Boot 2003).

Evaluating the battle of Mazar-i-Sharif, Rumsfeld (2002a) concluded that for the future U.S. military transformation, "[t]he ability of forces to communicate and operate seamlessly on the battlefield will be critical to our success. In Afghanistan, we saw composite teams of U.S. special forces on the ground, working with Navy, Air Force and Marine pilots in the sky, to identify targets, communicate targeting information and coordinate the timing of strikes with devastating consequences for the enemy."

The offensive against the Taliban regime was launched in late 2001.[15] Strategically, Operation Enduring Freedom (OEF) relied on a comparatively small force that was deployed rapidly to the other side of the globe. In operational terms, the quick victory was closely linked to air power. Troop communication crossed service lines, the Air Force and Naval aircraft provided close air support to special operation forces on the ground and dropped GPS or laser-guided precision bombs (mostly Joint Direct Attack Munition [JDAM]). Special operation forces were, "equipped with laser rangefinders, laser target designators, laptop computers and modern ultra-high frequency (UHF) radios. Air Force combat controller used the (...) laser rangefinder and designator. The system, which looked like a giant pair of elongated binoculars mounted on a small tripod, shot out a laser beam that allowed soldiers to calculate the coordinates of Taliban or Al Qaeda positions and call in air strikes" (Mahnken 2008: 198). It is important to note that critics doubted the conclusion drawn by the RMA-proponents and argued that (a) there was a lot more conventional fighting in Afghanistan than the transformationists wanted the public to believe, and (b) that the circumstances in Afghanistan were so special that it was problematic to adopt for any other case (Biddle 2003; Kagan 2006).

The quick victory was further attributed to the use of reconnaissance planes and drones and to combat drones that carried Hellfire missiles and engaged in the elimina-

---

15 For an account of the offensive in 2001 and the ongoing campaign since 2002 from the perspective of the U.S. Army, see "A Different Kind of War" (Wright 2010).

tion of targets. Thus, OEF was labelled as "a war to change all wars" (Arquilla 2008: 41). For some, it was a "sneak preview of future possibilities" (Arquilla 2008: 8), while for others it was a "showcase" of the transformation of the American military (Boot 2003). In his second speech to The Citadel, then President of the United States Bush implicitly pointed at the validation of NCW theory when he acknowledged that "commanders are gaining a real-time picture of the entire battlefield and are able to get targeting information from sensor to shooter almost instantly" (Bush 2001). Backed with these and other examples on how technology helped the Coalition forces to defeat the Taliban fighters, he drew the conclusion that the need to transform the U.S. military towards this technologically advanced 'future force' now was more urgent than ever. The campaign in Afghanistan led many in the military and the administration to believe the armed forces have to become less heavy, more agile and more expeditionary (Whittle 2001). The initial reluctance to follow Rumsfeld's transformation idea seemed overcome.

Learning from Afghanistan, transformation proponents thought that their assumptions had been confirmed, hence the forces deployed to Iraq were comparatively light (Weinraub/Shanker 2003). Initially, the invasion of Iraq in the spring of 2003 was similarly perceived as a "new American way of war" (Boot 2003). For many observers, the invasion of Iraq proved an initial success of the U.S. military transformation project as the outnumbered, yet technologically advanced coalition forces swept away the Iraqi military in only three weeks (Schreer 2003). According to an early account by Donald Rumsfeld, Operation Iraqi Freedom (OIF) was successful, as it combined speed, jointness, intelligence and precision (Rumsfeld 2003). Other assessments were less enthusiastic, as they found out that, in OEF and later in OIF, NCW was used only on a limited scale. Nevertheless, they still saw NCW becoming a reality (Wilson 2007): Examples for the unprecedented use of C4I during OIF included Predator and Global Hawk drones that were remotely operated via satellite links from the territory of the U.S.

The U.S. military benefited even further from enhanced situational awareness, as the U.S. Army Blue Force Tracking System Force XXI Battle Command Brigade and Below (FBCB2) was deployed. It comprised of GPS transponders that transmitted information about their location via satellites. Thus, a report from the U.S. Army War College concluded that, during OIF, U.S. forces were provided with a common operational picture and achieved satisfying situational awareness (Murphy 2005). Due to this improved situational awareness, deaths attributed to fratricide dropped from 24 percent to 11 percent compared to the first Gulf War. NCW was also praised for enhancing combat support and casualty tracking (Hawk 2005).

Analysing after-actions reports from the invasion of Iraq revealed, however, that, due to bandwidth restrictions, gaps in equipment and the breakdown of the communication network in severe weather conditions, a resilient, real-time operational picture was widely absent (Schwiebert 2004). In the volume "On Point" – the U.S. Army Chief of Staff's campaign history of OIF – the overall assessment of NCW was balanced (Fontenot et al. 2005: 416): "The Army's investment in digitization paid off in OIF, or rather showed promise. Army units fought enabled by a digital network that allowed them to see their units and their activities, which led to situational understanding. Confident that they knew the location of their units, commanders could decide rapidly where, when, and how they would be employed. Additionally, because of joint initiatives in communications and networking and the provision of selected Army systems to other services, coalition ground forces could fight joint net-enabled operations."

Yet, interoperability remained a problem. The report concluded that the U.S. Army might not have been "a truly netcentric force (...) [but rather] a net-enabled force, one that was significantly more effective because of digitization efforts since Desert Storm" (Fontenot et al. 2005: 418).

When the coalition forces remained in Afghanistan and Iraq after the initial invasion/combat phase, critics started to voice doubt whether the RMA-inspired U.S. military transformation could bring about an Army able to deal with low-intensity conflict situations and insurgency. The growing number of U.S. casualties to car bombs, roadside bombs or mortars and rocket-propelled grenades fuelled the debate between proponents and opponents of the U.S. high-technology transformation and – more specifically – NCW (Talbot 2004). When Donald Rumsfeld resigned over the criticism of his course of action in Iraq, the technology-driven military transformation that he had pushed came to a halt. As *transformation* and *NCW* had been so heavily intertwined, the inevitable question was: What had become of NCW?

2.7 Is NCW dead?

By the end of 2006, things were different compared to 2001:[16] The two most powerful and dedicated supporters of NCW, Donald Rumsfeld and Arthur Cebrowski, had left the Pentagon, the Office of Force Transformation had been dissolved, the Army's Future Combat System was considerably downsized by the new Secretary of Defense and the technology-driven transformation project was discredited. In summer 2010, Secretary of Defense Robert Gates had recommended eliminating the USJFCOM – the U.S. military transformation command. Gates insinuated that the goal of USJFCOM – to strengthen jointness among the armed services – had been achieved (Garamone 2010).

Unsurprisingly, as NCW was one of the core concepts of the U.S. military transformation project, the concept came under pressure and, in the end, vanished together with transformation – at least as a term. As a comparison of the 2006 and 2010 QDR shows, in 2006, the official view was that "recent operational experiences in Afghanistan and Iraq have demonstrated the value of net-centric operations" and that the Department sought to increase net-centricity (Rumsfeld 2006: 58f.). Yet, the terms "network-centric" and "net-centric" completely disappeared from official Pentagon publications after 2006 (Shachtman 2007) and, in the QDR 2010, there is no mentioning of NCW (Gates 2010). Suffice it to say that the term transformation did not appear either.

Yet, it might be premature to conclude that, with the term NCW, the idea of NCW has also vanished. The basic technological elements of NCW were still considered important. The 2010 QDR (Gates 2010) concluded, for example, that "U.S. forces would be able to perform their missions more effectively – both in the near-term and against future adversaries – if they had more and better key enabling capabilities at their disposal. These enablers include rotary-wing aircraft, unmanned aircraft systems, intelligence analysis and foreign language expertise, and tactical communications networks for ongoing operations, as well as more robust space-based assets, more effective electronic attack systems, more resilient base infrastructure, and other assets essential for effective operations against future adversaries."

16 This sub-section headline is inspired by the title of Elizabeth Quintana's article *Is NEC dead*? in which she assessed the state of the British NEC concept in 2007 (Quintana 2007).

This quote shows that the awareness for information networks, advanced C2 and unmanned systems has not decreased since the U.S. military transformation project dried up; these technologies will likely continue to play an important role in future procurement decisions and organisational adaptation. Furthermore, within the individual services, the network approach is still considered. Yet, the emphasis had moved away from the network to the commander. Network-*centric* today had become network-*enabled*.[17] It would be appropriate to conclude that NCW as a sound concept has lost the accentuated status it had between 2001 and 2006 and now is one defence concept among others, such as counterinsurgency (COIN), cyber security and civil-military cooperation.

2.8 Critical Voices on Network-Centric Warfare

So far this chapter traced the intellectual development of the NCW concept, its implementation into the American military as well as its downturn in the face of the actual testing of the concept's promises in the wake of the Afghanistan and Iraq Wars. The decline of the NCW concept's political support was matched by a plethora of criticism on various aspects of the NCW concept. The most important lines of criticism will be discussed in the remainder of this chapter.

Basically, two strands of critique developed since the concept was issued in 1999. On a broader, more strategic or even philosophical level, the NCW concept was questioned as such. Some critics, for example, said that NCW misunderstood the very nature of war. Other distinguished strategists and observers of U.S. defence policy argued that the concept was unfit for the present security challenges such as asymmetric warfare and insurgency. Apart from these general disapprovals, another strand of criticism evolved that in principle agreed on the merits of NCW, but critically pointed to single aspects of NCW or on the chosen path of concept implementation in the U.S. Some of these limitations have been acknowledged by the writers of *Network-Centric Warfare* themselves (Alberts et al. 1999: 5f.).

2.8.1 General Lines of Criticism

Alfred Kaufman (2002) argued that NCW was ignoring the human dimension in war. He was particularly critical of the NCW concept's stress of accuracy and real-time action that ignored the potential counter-measures of the adversary on the one hand, but also the limited capacities of one's own military organisation on the other (also see Kaufman/Welch 2004). He furthermore disagreed with the assumption of the advocates of NCW that such challenges can somehow be overcome (Kaufman 2002: 21): "The belief in the absolute perfectibility of things expressed in [the assumption of perfect accuracy and zero time delays, I.W.] has the alluring sound of all -isms that populate the graveyard of history; it does violence to reality by subjugating it to an utopian vision from the vantage point of which all natural limits become mere obstacles (...)."

---

17 No official document on the adapted NCW concept had been issued by the services (personal conversation with a U.S. Army Lieutenant Colonel form the 5th U.S. Signal Command, USAEUR, Oberammergau, 5 November 2010). For a mentioning of the new approach see the contribution of Gen Mattis, commander of the USJFCOM (Mattis 2009).

Such "mere obstacles" included, for example, the need to train new types of soldiers with the necessary technical skills or the precondition of air superiority in a theatre of war where drones and manned reconnaissance aircraft are expected to deliver the information needed to gain information dominance (on the latter point see Wilson 2007: 26). Kaufman made another related point: The integration of all platforms into one network might just create an Achilles Heel. "In any event", he argued, "by thus concentrating our capabilities into one ecosystem we invite the enemy in one blow what otherwise may have required many" (Kaufman 2002: 22). He goes on by saying that in a network-centric organisation, weapon systems will become "virtually dependent on the life-giving information flowing through the embedding organism. Terminate that, and you will no longer have the collection of independently capable weapon platforms to fall back on, but a bunch of dead cells. That is a recipe for disaster" (Kaufman 2002: 23). This point of criticism did not go unheard: In 2008 the USJFCOM warned about creating a vulnerability by becoming too network-dependent (USJFCOM 2008: 23).

Elsewhere, Kaufman and Welch furthermore argued that NCW theory had not been verified and had been taken over too hastily by the U.S. DoD (Kaufman/Welch 2004). They made the argument that the appearance of Network-Centric Warfare theory and the speed by which it has been made the centre of U.S. Armed Forces' transformation had put long-hauled principles of systems analysis for defence planning at odds. Developed in the Pentagon in the 1960s, Systems Analysis is a planning approach that used intellectual discourse and scientific, quantitative methods to derive military decisions.

Other scholars also argued that NCW theory was unfounded and pointed at a variety of assumptions made without proof (Reid et al. 2005; Storr 2009; Vego 2007). Another strand of criticism did not necessarily question the usefulness of NCW as such, but argued that the concept was not fit for meeting the current security challenges presented to Western democratic states. NCW, it is argued, had been issued as a new and revolutionary concept at a time of a strategic vacuum (that is, between the end of the Cold War and the terror attacks of 11 September 2001). The concept would resemble the broader trend after the Cold War to do capability-based planning instead of threat-based planning. Looking at NCW from that angle, it could just have been another rhetorical back door for the U.S. Armed Services to retain large defence budgets for acquiring up-to-date technologies (Hammes 2004; Kaufman 2005). While this fact in itself might not contradict the operational value of NCW, the critics go on to question the concept on more substantial grounds.

Thomas Hammes, a retired U.S. Marine officer, argued in *The Sling and the Stone* that NCW was ill-suited to cope with what he calls 4th Generation Warfare (4GW) (Hammes 2004).[18] Over the centuries, he explained, warfare developed from massed direct-fire weapons (1GW) to defensive trench warfare (2GW) to manoeuvre warfare (3GW) to an evolved form of insurgency (4GW), which Hammes defined as follows (Hammes 2004: 2): "Fourth-generation warfare (4GW) uses all available networks – political, economic, social, and military – to convince the enemy's political decision-makers that their strategic goals are either unachievable or too costly for the perceived benefit. It is an evolved form of insurgency, still rooted in the fundamental precept that superior political will, when properly employed, can defeat greater economic and military power, 4GW makes use of the society's networks to carry on its fight."

18 For a critical discussion of 4GW see Junio (2009).

Hammes did not believe that the current U.S. Armed Forces transformation enabled the U.S. to wage war against a 4GW enemy (Hammes 2004: 205): "In sum, Information-Age technology has actually eroded our position relative to potential 4GW enemies over the last couple of decades. (...) Although we have technology superior to that of many of our potential adversaries, we continue to use it to support a third-generation style of warfare. In contrast, the most forward-thinking of our opponents have seized on fourth-generation warfare."

Specifically with regards to NCW, Hammes claimed that the current information systems are focussed on finding conventional forces which the enemies of today are not: "The target is fundamentally different from what our systems and organizations were designed to collect against" (Hammes 2004: 194). Elsewhere in his book, Hammes also argued that NCW was too focussed on the tactical level while strategic considerations were left out. The new DoD concepts, he said, were "about winning battles, not winning wars" (Hammes 2004: 225).

Colin Gray makes a similar argument. In publications issued after the post-invasion phase in Iraq, Gray warned that technology must not compensate "for unwise political vision, faulty high policy, or erroneous strategy" (Gray 2006c: 47). Although America is usually superior at making war, it is far less superior in making peace out of war. Gray argues that the current military transformation, though certainly welcome, cannot correct the long-standing U.S. weakness in the proper use of force as an instrument of policy. "There is a lot more to war than the operational level", he concludes elsewhere (Gray 2005). In his view, the American way of war is characterised by a number of attributes and technology-dependence is one quality among those that Gray ascribed to U.S. war fighting.[19] He acknowledged a tension between a culturally fixed American way of war and the real-world challenges of irregular wars and terrorism. Like Hammes, Gray is convinced that the "technology-dependent, stand-off style [of war, I.W.] is not appropriate for the conduct of war against irregulars (...)" (Gray 2006a: 49).

Similarly, after having analysed the U.S. military operation in Iraq and the problems of bringing peace to the country after the successful invasion-phase, Williamson Murray and Robert Scale remark (2003: 182): "For all the effects-based operations and operational net assessment, the failure to understand the enemy where he lives – his culture, his values, his political system – quickly leads up a dark path where any assumption will do." They conclude "[t]echnology is useful in conventional warfare, but not decisive in the new kind of conflicts. (…) The enemy will be located not by satellites and UAVs but by patient intelligence work, back-alley payoffs, information collected from captured documents, and threats of one-way vacations to Cuba" (Murray/Scale 2003: 236–237).

An example of how NCW is ill-equipped for urban warfare is the over-reliance on sensor-based situational awareness, which may not fully reflect what is going in the urban theatre (Wilson 2007: 8). Clay Wilson mentioned another possible drawback of data-dependent military organisations. Under the heading "perverse consequences", he lists various likely problems: The efforts it takes to further invest in networking to the lowest level against the background of a possible diminishing marginal utility (economies of scale), shifts of mission objectives due to information-superiority, and manage-

19 According to Gray, the American way of war is apolitical, a-strategic, a-historical, problem-solving/optimistic, culturally-challenged, focused on firepower, large-scale, aggressive, offensive, profoundly regular, impatient, logistically excellent, and highly sensitive to casualties.

ment overconfidence due to the reliance on information systems (Wilson 2007: 49ff.). Stephen Biddle approached the value of NCW from a different, exogenous angle. By analysing the military campaigns, he cast doubt about the NCW success stories in Afghanistan and Iraq. He conducted interviews with a broad range of key American participants in the war, analysed available documents and concluded that the war in Afghanistan as a whole "was much more orthodox, and much less revolutionary, than most now believe" (Biddle 2003).

In the case of Iraq, he made the claim that the coalition's success was in fact not entirely due to the smooth interaction of networked weapons and soldiers. Iraqi armed forces simply made it all too easy for the U.S. troops to occupy and control their territory by destroying the infrastructure in order to prevent U.S. forces from moving quickly and not concealing themselves properly. In short, the Iraqi military made many tactical mistakes; soldiers were badly trained on the use of their weapons as well as in urban warfare. It was not only the coalition's strengths that lead to a quick success but also the shortcomings of the Iraqi military. Inferring from the success of the U.S. in Iraq that the network-centric way of waging war will win the next war would therefore be dangerous (Biddle 2007). David Betz, moreover, stated a number of occurrences in both the Iraq and the Afghanistan campaigns in which NCW on the side of the Alliances actually failed to work (Betz 2006).

Milan Vego argued in a similar fashion as Biddle. He doubted the effectiveness of NCW in the face of a strong, resourceful enemy given the fact that empirical evidence so far had only shown its effectiveness in fighting the weak and passive Taliban regime in Afghanistan or the military in Iraq. Vego further asserted that surgical operations with as little exposure as possible to the enemy obscured the strategic importance of gaining and holding actual control of the enemy's territory. Another point of criticism concerned tactical-level jointness; Vego doubted that jointness below the operational level would be useful as service cultures, doctrines and command processes, which were simply too different to force them into a joint organisation (Vego 2007).

Finally, Frederick Kagan, in his critique of the overall U.S. American military transformation, not only found that the NCW advocates were largely ignorant of already existing military transformation efforts, he argued that they were engaged in an exercise of "reinventing the wheel" by presenting NCW as a new capability and a new set of requirements when, in fact, it was largely a logical progression from efforts already well underway since the 1990s (Kagan 2006). Even more importantly, he claimed that, in the strategic pause during the 1990s, an inwardly-focused transformation started, looking only at one's own capabilities and programmes. NCW visionaries gathered much of their 'proof' from the business world and not from strategic threat or risk assessments (Kagan 2006: 213).

2.8.2 Specific Issues

Apart from these rather general assessments of the overall usefulness of NCW, there were also more specific observations on parts of the NCW concept or on problems experienced during implementation that will be discussed in the remainder of this section. These issues comprise of questions of network resilience, software vulnerability, bandwidth, the interoperability of systems, information overload and micromanagement, as well as problems of sufficient training, challenges posed to the acquisition system and the open access of U.S. networking plans.

The first problem is that of physical network resilience. The network can either be physically disrupted by the adversary through mechanically attacking the network components (shooting satellites down) or through preventing data transfer by using electromagnetic interference (jamming). While, for Kaufman, this flaw disregards the NCW concept as a whole, others address counter measures against jamming or satellite security. Software vulnerability is a related issue. In 2008 and 2009, the Conficker computer virus infected computers in the British, German and the French Defence Ministries and even in French fighter aircrafts.[20] As a reaction, the U.S. and Germany, for example, set up military cyber security units in 2009.[21] Bandwidth and sufficient data transfer capabilities are another problematic issue in current operations. As their own capabilities are not sufficient, the U.S. and other armed forces have satellite capabilities that need lease bandwidth from commercial providers, which might have drawbacks in terms of encryption standards (Wilson 2007: 16ff.). Furthermore, in a joint or multilateral operational context, the issue of interoperability appears. Early adopters of NCW, such as the U.S. or the UK, have to take into account that potential coalition partners might not have up-to-date C2 equipment.

Apart from the purely technical aspect of interoperability, the issue also has a cultural dimension. Soldiers with different cultural backgrounds (such as an Army or an Air Force member, a German or a Japanese military officer) might interpret one and the same information (acronyms/slang, colloquialisms) differently. In a coalition network, this could cause deadly mistakes. Merindol and Versailles give a concrete example of such a situation: In a combined training exercise, a UK Forward Air Controller (FAC) was directing a U.S. pilot. The British solider tried to direct him to a road between two different coloured fields. The U.S. pilot reported to have clear contact with the "dirt ball road". The UK FAC did not understand what the U.S. pilot meant (Merindol/Versailles 2007). Another good example is silence. In some military cultures silence is understood as a confirmation, in others, as a rejection.[22] Another such human-related problem to achieving NCW is the desirability of the promises NCW makes (information dominance and self-synchronisation) or the limits of the human capacity to exploit these. There is only a thin line between information dominance and information overload (Brown 2001). Furthermore, a simple information advantage does not necessarily have to lead to a better understanding of the situation. David Betz illustrated that point by referring to Arquilla's and Ronfeldt's (Arquilla/Ronfeldt 1997 (1993): 24f.) chessboard metaphor (where you see the entire board, but your opponent sees only his or her own pieces which would give you an advantage): "If Gary Kasparov and I both look at the same chessboard he will understand much more of what it means than I because he has vastly greater training and skill" (Betz 2006: 520).

Self-synchronisation, furthermore, could easily lead to aimless or incoherent behaviour (Alberts/Hayes 2003: 184). Especially in Germany, it was discussed whether NCW

20 See http://www.telegraph.co.uk/news/worldnews/europe/france/4547649/French-fighter-planes-grounded-by-computer-virus.html and http://www.theregister.co.uk/2009/01/20/mod_malware_still_going_strong.

21 The German Bundeswehr assembled a Cyberwar unit in February 2009. The unit called "Computer network operations and information centre" is subordinate to the Bundeswehr Strategic Reconnaissance Centre. Similarly, U.S. Defense Secretary Robert Gates approved of the creation of a unified U.S. Cyber Command (USCYBERCOM) to oversee the protection of military networks against cyber threats in June 2009.

22 I am grateful to Professor Trevor Taylor of Cranfield University for providing this example.

would undermine the principle of mission order tactics so highly valued in the German command style and instead open the door to micromanagement by the hierarchically higher levels (Mey/Krüger 2003). Sufficient training appeared to be a problem as well, as Stuart Griffin and David Whetham point out in their case study on the application of Network-Centric Operations by the UK in Iraq (Griffin/Whetham 2007). The same problem has been observed among U.S. soldiers who rather relied on (non-secure) emails and chat-rooms instead of the military-only decision-support systems to which the soldiers were not familiar (Wilson 2007: 25). Furthermore, a network-centric military must attract people with the necessary IT skills to maintain the network and develop it further.

Finally, the open style of discussion about the NCW concept and the adoption of NCO as one pillar of the U.S. transformation raised concerns in the U.S. that this could encourage potential adversaries to explore counter measures to networked operations (Wilson 2007).

Naturally, the promoters of NCW theory were not ignorant of the different lines of criticism raised. In a rejoinder to some of the issues concerning the validity of the RMA argument and the role of technology in war that were debated, Cebrowski said that "none of us [the revolutionaries] denigrated military experience, training, or reliance on 'humint' – information from human sources – whether they were spies or soldiers on the battlefield. Nor did we think in the absolute terms the critics accused us of doing. On the contrary, as a group, the revolutionaries in the Pentagon during the 1990s focused on probabilities, not dogma" (quoted in Blaker 2007: 46).

More specifically, Cebrowski admitted that NCW was not likely to eliminate the fog of war – an assumption that opponents to NCW were eager to prove wrong – but at least to reduce it considerably. In the case of asymmetric threats and suicide attacks, Cebrowski argued in a similar way (Blaker 2007: 60f.): "While NCW might not be the only and not the best solution, it nevertheless can contribute to better protection, as suicide bombers could be identified earlier and eliminated without collateral damage."

In sum, a vivid and widespread debate about NCW paralleled the implementation of the concept. In 2006, against the background of increasing criticism Colin Gray remarked that the transformationists "have lost the debate, though whether they lose in the shaping of the process of U.S. military transformation is, of course, another matter entirely" (Gray 2006b: 9).

2.9 Conclusion

Between 2003 and 2005, the Department introduced NCW widely into the U.S. military through the means of centralised oversight. Despite a plethora of general and substantial criticism on NCW, the network idea dominated U.S. military change in the early years of the new century.

Today, NCW has lost much of its prominence. Many presumably equated NCW with the technology-driven military transformation of Donald Rumsfeld. When his transformation vision was discredited during the Iraq war, NCW also came into the line of fire. While the importance of information technology for the conduct of military operations is still widely acknowledged, NCW has lost its prominence. Yet, the concept had had a significant impact inside and outside the U.S. military. Reports from military operations today might not speak of network-centric operations. They are present nevertheless – for example in the remotely operated U.S. drones in Afghanistan, Pakistan and

Yemen. Some of the ideas put forward by Arthur Cebrowski and his co-writers in the beginning of this century have survived the apparent demise of the NCW concept.

3 Military Concept Adoption

The main theoretical argument developed on the following pages is that the variance in NCW concept adoption between the UK and Germany can be explained by the differences in relevance of rational and institutional drivers in the two cases of adoption. The chapter starts by answering the question: "What is this a case of?" (Levy 2008: 2). The spread of NCW from the U.S. to other military organisations is a case of concept diffusion and adoption. The chapter proceeds by focusing more deeply on the phenomenon of concept adoption. Categories like 'adoption pace' and 'adoption patterns' are developed to empirically grasp the differences in the introduction of NCW in the UK and in Germany. An analytical framework is presented that combines rational and institutional explanations, which can account for the empirically observable differences in NCW concept adoption in the UK and Germany.

3.1 Diffusion, Dissemination and Adoption

Despite the existence of a few comparative studies on the spread of information technologies in military organisations (Adams/Ben-Ari 2006; Demchak 2003; Terriff et al. 2010) theoretically-guided, in-depth, comparative case studies on the subject are still rare. This rarity does not mean that there is a general lack of comparative research on the introduction of new ways of warfare. Indeed, in the field of strategic studies there is a vast literature on the spread of doctrines or technologies among military organisations. These studies employ different explanatory models tracing military innovation and diffusion by focusing on cultural, systemic or institutional factors.

Eitan Shamir, for example, analysed the introduction of the German principle of mission order tactics into the British, U.S. and Israeli militaries using an organisational culture framework (Shamir 2008). Dima Adamsky developed a strategic cultural framework to explain the different approaches of debating the impact of information technology in the Soviet, American and Israeli defence communities (Adamsky 2010). João Resende-Santos (2007) gave a careful account of the changes in South American armies at the end of the 19th and the beginning of the 20th centuries. Using a neo-realist framework, Resende-Santos explained how Chile, Argentina and Brazil imported the German and later the French models of military organisation. Emily Goldman (2003) found that sociological institutionalism would provide the best tools to analyse the differences in introducing aircraft carriers into the navies of Britain, the U.S., Japan, Germany and Italy in the 1940s. Focusing less on military doctrines or technologies (understood as ideas) but on the proliferation of (tangible) conventional weapons, Eyre and Suchman also employed institutionalist reasoning (i.e. the role of high- and low-symbolic significance weapons) to complement existing rationalist explanations (notably superpower manipulation, national security and factional interests) to explain the diffusion of weapons (Eyre/Suchman 1996). Against this background of existing theoretical foundations on military diffusion, studying the introduction of NCW into the British and German militaries can thus benefit from the methodology of empirical work and theoretical reasoning in those and other comparative studies.

3.2 Diffusion of Innovations

How could the fact that the UK and Germany (as well as other states) introduced a foreign concept that was invented by the U.S. best be labelled? What is it a case of? One could argue it is a case of *military change* in general, or, more specifically, one could speak of *military innovation*. Terms like military diffusion, emulation, adoption and adaptation are also quite commonly used by other scholars and are relevant categories for studying the introduction of a foreign concept by a military organisation (Eliason/Goldman 2003; Farrell/Terriff 2002b; Resende-Santos 2007). It is, however, worth digging a bit deeper here and looking for a more accurate categorisation.

Applying a global or system's view, the phenomenon under study – the spread of a foreign concept and its adoption in organisations or states – is a case of diffusion. In the organisational literature, diffusion is the communication of information about an innovation and its adoption (Eliason/Goldman 2003: 12). This narrow definition centres on one defined technology or concept. A richer definition is offered by the literature of comparative politics in which diffusion refers to a process where "policy decisions in a given country are systematically conditioned by prior policy choices made in others" (Simmons et al. 2006: 787).

In a nutshell, diffusion is the "social process by which an innovation spreads through a social system over time" (Webster 1971: 178). The initial conceptualisation and examination of the innovation diffusion phenomenon were made in the field of rural sociology when sociologists conducted pioneering work on the diffusion of agricultural innovation such as high-yielding hybrid seed corn, chemical fertilisers and weed sprays (Singhal 2002). Everett M. Rogers, at the time a doctoral student with a family background in farming studied the diffusion of weed spray in Iowa farm communities (Rogers 1960). His interest in the general patterns of diffusion prevailed and his seminal work *Diffusion of Innovations*, published in 1962, laid the groundwork for new research agendas not only in rural sociology but also in a number of other disciplines. In fact, research on the existing diffusion literature in the late 1960s revealed that empirical and theoretical work on diffusion covered disciplines as diverse as psychology, anthropology, economics, communications, public education, journalism and medical sociology (Rogers 1968; Rogers et al. 1967). A variety of organisations became subjects of diffusion research, notably commercial firms, hospitals, universities and schools, but also social groups and eventually the social systems in which diffusion occurred (national or global industries, the health sector, etc.). Depending on the research interest, the innovation whose diffusion was studied either was material (a new drug, a new weed or a new machine) or immaterial (concepts, business procedures, programmes or organisational structuring). Later, diffusion was considered a useful theoretical concept in public policy research that could illuminate the spread of economic and political liberalism across countries (Simmons et al. 2006) or the spread of national welfare policies (Thomas/Lauderdale 1988). Studies in international relations and international security, most prominently, focused on the diffusion of international norms (Checkel 1999; Finnemore/Sikkink 1998; Jepperson et al. 1996; Price/Tannenwald 1996). Finally, since the 1990s, considerable work has been done on diffusion theory, especially on military diffusion (Goldman/Eliason 2003; Terriff et al. 2010) and recently diffusion theory has even been used to explain the spread of suicide terrorism (Horowitz 2010b).[23]

50 years after its initial publication, *Diffusion of Innovations* – the first edition appeared in 1962, the fifth in 2003 – is still a classic standard work of diffusion theory and believed to be the second most cited book in the social sciences (Singhal 2002). In *Diffusion of Innovations* Rogers defined and categorised several aspects of diffusion. He broke down the diffusion process into different phases, theorised about the nature of innovators, early adopters and laggards and about their numerical allocation within a social system. He was, furthermore, concerned with uncovering the social communication processes involved in diffusion by introducing the concept of opinion leadership in diffusion processes. Rogers' work has inspired and guided extensive empirical research throughout various disciplines: Some research focused on the different types of adopters (Mahajan et al. 1990), whereas other scholars studied the patterns of the timing between the occurrence of an innovation and its adoption (Sarvary et al. 2000).

The research design of this book benefited from definitions and categorisations from the diffusion literature. Building on these works, the following sections introduce categories of diffusion theory that helped in the operationalisation of the research argument.

3.3 Diffusion Processes: Dissemination and Adoption

A closer look at Rogers' definition of diffusion reveals that he adopted a system's view to describe the diffusion phenomenon. He defined diffusion as "the process by which an innovation is communicated through certain channels over time among the members of a social system. It is a special type of communication, in that the messages are concerned with new ideas" (Rogers 2003: 6). Furthermore, diffusion "is a kind of social change, defined as the process by which alteration occurs in the structure and function of a social system" (Rogers 2003: 7).

Quantitative research on diffusion applies this system's view and tries to specify the spread of an innovation throughout the whole system and to come to a reasonable conclusion about the timing concerning when an innovation finally reached a certain saturation level within the system. The relative speed with which the innovation is adopted in the whole system usually resembles an S-curve (Rogers 2003: 25). In the area of military diffusion, Demchak's study on the introduction of information technologies in military organisations would be an example of a quantitative systems-level research design (Demchak 2003).

Research on diffusion can also adopt unit-level perspectives and therefore allow for different research agendas. At the unit level there are two complementary diffusion processes: On the one hand, there is the *dissemination* process of the exporting unit. Here, research is focused on the role of the innovator as an opinion leader within the system (Baumgarten 1975). With a view to the spread of the military transformation idea from the U.S. to other NATO states, Farrell and Terriff also point to institutional structures such as international organisations and policy networks to promote, transmit and sustain the transformation idea (Farrell/Terriff 2010).

On the other hand, there is the actual process of introducing an innovation by the importing units. This phenomenon will be labelled as adoption. In the literature, the

23 The Joint Center for International and Security Studies (JCISS) launched project on the "Diffusion of Military Innovation" in 1997. See http://www.au.af.mil/au/awc/awcgate/innovation/jciss/diffusmilinnov.htm (accessed 28 January 2010).

adoption process has also been called adaptation, borrowing, emulation, imitation or copying. Yet these terms carry a certain qualitative connotation with regards to the charac-teristics of the adoption process and should therefore rather be used to distinguish different adoption patterns than to be used synonymously with adoption. The duality of dissemination and adoption was captured in the marketing literature by pointing at the "supply and the demand" side of diffusion (Brown 1981: 1–14). In public policy research, these two perspectives on diffusion had been referred to as the "push and pull" of policy-relevant ideas (Majone 1991: 104).

Depending on the research question, empirical work either aims at describing or explaining dissemination or adoption.[24] Research on the dissemination process, for example, might ask if and how innovators or opinion leaders actively tried to spread a technology or concept.[25] If such a view were applied to explain the spread of NCW, the research focus would be on the U.S. military and the role of policy-makers from the Pentagon, military officers and the defence industry in distributing the concept (for a discussion of the industry's role in pushing transformation concepts see (Dombrowski/ Gholz 2006)). Comparisons could be made either with other cases of earlier military innovations or with contemporary cases of civilian U.S. innovations.

3.4 Adoption

This book, however, does not so much focus on the U.S. as the inventor of NCW, but on the armed forces of Germany and the UK – two military organisations that adopted the concept of NCW. As a consequence, the introduction of the concept of NCW into the military organisations of the UK and Germany is a case of *adoption*. This is not to say that dissemination does not matter as the dissemination strategy of the U.S. might, in one or in both cases, have influenced the path of NCW introduction.

Adoption is the process by which an organisation introduces a foreign concept. For each respective organisation, adoption of a foreign concept results in the change of its structures, rules and routines. The adoption process can be broken down into its subprocesses. This is useful to structure empirical research on adoption. Collin J. Bennett, for example, breaks the path of adoption of foreign policy programmes down to three stages. "Awareness of that program, utilization of that knowledge and adoption of the same program", he states, "are three conceptually and empirically distinct processes" (Bennett 1991: 33). These categories are useful for structuring empirical research as they allow for a better comparison between different cases of adoption with regards to timing, actors involved, outcomes and processes. Rogers offered a flow-model that included the five stages knowledge, persuasion, decision, implementation and confirmation (Rogers 2003: 22).This approach can be criticised, however, because these categories may be more suitable for tracing the adoption of a new technology rather than an intangible concept as it omits the phase of potential re-conceptualisation of the innovation. As NCW is a new concept and therefore potentially subject to being re-forged in

24 The conceptual boundaries around dissemination and adoption, around senders and receivers of innovations have been eased elsewhere. Those boundaries are either blurred through agents acting in between the sender and the receiver (Swan et al. 1999) or the receiver might surpass the sender in the exploitation of the innovation and diffuse practices of innovation adaptation back to the original innovator (Goldman 2003: 281).

25 Rogers developed the concept of change agents (Rogers 1995) that have an interest in spreading new technologies or ideas. These need not necessarily be the original innovators.

the process of adoption, for the purpose of this study, the adoption process of NCW in the UK and Germany is divided into the three temporal stages of (1) knowledge acquirement, (2) utilisation and (3) implementation. This compartmentalisation of the process, of course, is not rigid. In reality, situations of overlap might occur.

3.4.1 Adoption Phases and Stages

The first phase, knowledge acquirement, covers three sub-stages. Knowledge acquirement starts with the (voluntary or involuntary) dissemination of the primary information about the innovation by the inventor (or opinion leader) until the adopter becomes aware of the innovation. This proceeds with a decision to actively acquire further information on the innovation. Finally, the first phase encompasses the actual search for the information. The second phase is utilisation, which also encompasses three stages. Yet, these stages do not necessarily have to be in the order presented here, as the respective processes can occur in parallel. One stage is the re-conceptualisation of the innovation in order to make it fit the organisation's own purpose, structure and processes. Even though the introduction of a foreign new concept is most likely to change existing organisational features, this is not a one-way street, but rather a mutual adaptation and the new innovation is also likely to undergo some change. This stage is supposedly more important when it comes to adopting an immaterial innovation such as a concept or programme. However, in the case of material innovations (a new machine or an information technology), certain adaptations of organisational routines might be necessary. Therefore, the term adaptation should be used here. A second possible stage is the legitimation of the introduction of the new concept. The organisation will use communication processes to explain (and, under certain circumstances, defend) the introduction of the new invention to its members and the external environment. A third possible stage is the actual decision to adopt the innovation and the decision-making process before that decision.

The final phase of the adoption process is the implementation of the (adapted) innovation and can also be divided into three stages: The initial stage, the transitional stage and the final stage. Depending on the nature of the innovation, the initial stage comprises of further conceptual re-adjustment of the concept and potentially of acquisition decisions, and the purchase of new technologies, the decision to re-structure the organisation and changes in organisational procedures. In the transitional stage, new equipment is delivered and fitted into the existing structures. Organisational routines begin to change. Moreover, re-conceptualisation of the innovation might take place as implementation goes on. This re-conceptualisation can be triggered by internal or external events and can take the form of further adjustments, but it can also include radical resetting. Finally, the introduction of the new innovation is regarded by the organisation as being complete.

Table 3: Adoption Phases and Stages

Adoption Phases	Adoption Stages
Knowledge Acquirement	From Dissemination to Awareness From Awareness to Decision to Active Knowledge Acquirement Knowledge Acquirement
Utilisation	Adaptation Legitimising Decision-Making and Decision to Adopt
Implementation	Initial Stage Transitional Stage Final Stage

Dividing the adoption of NCW into its sub-processes should make process-tracing easier. Knowledge acquirement, utilisation and implementation are, however, not analytical comparative categories or – to use social sciences' terminology – dependent variables. Therefore, the next important question to ask is: What are the meaningful dependent variables that allow for an inference about the nature of the NCW adoption in Germany and the UK? The following section provides an answer to this question.

3.4.2 Adoption Pace and Adoption Patterns

Frederick Webster offers a good starting point for finding categories to compare cases of adoption. Analysing the existing diffusion literature (mainly in the field of marketing), Webster finds that research is either concerned with the adoption *outcome* or with the adoption *process* (Webster 1971: 179). Outcome-oriented research is concerned with the timing of the adoption, the investment and the consequences, the latter being changing costs, revenues and profits. Process-oriented research emphasises the social aspects of diffusion (Webster 1971: 179). Webster acknowledged that there was no absolute dichotomy between the two agendas and, in fact, a combination of both might enhance our understanding of diffusion processes. Therefore the comparison of NCW concept adoption in Germany and the UK will take into account both the adoption outcomes and the adoption processes.

Inspired by analytical frameworks found in the literature (Goldman/Ross 2003) in this book the adoption processes in the UK and Germany were studied with special attention on (1) *timing and pace*, (2) *concept faithfulness* (the differences between the original concept or technology and the adopted one) and (3) *faithfulness of implementation.*

The categories timing and pace consider the periods between the different phases and stages of adoption and are measured in months and years. As the case studies show, the German Bundeswehr was comparatively late and slow in acquiring knowledge on NCW and in adapting the concept. The British, on the contrary, actively tried to keep up with the U.S.' pace on NCW introduction.

Secondly, the concept faithfulness of the national version of NCW will be assessed. In the literature on diffusion, faithfulness is a notion to make assumptions on the degree to which the adopted concept resembles the original or – to change the perspective – to refer to the differences between the original concept or technology and the adopted one. (Jacoby 2004: 6f.). In the case of high faithfulness, the original concept and the adopted version are quite similar to one another. Less faithful adoption occurs when the introduced concept varies considerably from the original. Goldman and Ross distinguished between imitation and replication; adaptation and modification (changing the concept so as to fit it into the existing institutional patterns of the adopter), and selective emulation (Goldman/Ross 2003: 387).

Table 4: Modes of Concept Faithfulness

	High Faithfulness	Medium Faithfulness	Low Faithful
Adoption Mode	Copying	Modifying	Selecting
Adoption Outcome	Copy	Conversion	Template

The third category is faithfulness of implementation. Looking at the faithfulness of the implementation allows for conclusions about the relationship between the officially expressed intent to introduce NCW and its actual implementation (resources allocated, delivery of equipment and changing of procedures) to be drawn.

Initial research on the introduction of NCW in Germany and the UK showed a considerable difference in their timing, concept faithfulness and faithfulness of implementtation (see table 1 above). It appears that the UK was much quicker to embrace the new concept, less faithful in the concept adaptation, but in the end more faithful in fulfilling its own objectives. Germany, on the other hand, appeared to be rather slow in all of the three adoption phases. While the German NCW concept itself is more similar to the original NCW, it showed less faithfulness during implementation.

The diffusion literature has provided definitions and categorisations to distinguish the phenomenon that is studied here – adoption – from diffusion and dissemination. The diffusion literature has also helped to break down the adoption process. One might object that this division into three adoption phases is artificial and mechanistic.[26] Admittedly, this framework might limit the richness of the case studies and even might result in omitting otherwise interesting potential factors. The division, however, helps to reduce the complexity of the case studies, which allows for generalising about the phenomena observed. More importantly it allows for a structured comparison. This is important for tracing the processes by which NCW was introduced into the British and German military organisations and for answering questions such as: When, how, and with which institutional entities or individuals did NCW enter the military organisations observed? Who decided on the introduction? What differences exist between the original NCW conception and the national version? How were the concept and its introduction discussed and, finally, how (well) did the implementation of the national concept proceed? Although equipped with a parsimonious analytical toolkit, this undertaking is largely descriptive.

The second aim of this book is hence to *explain* why the UK has been more successful in introducing the NCW concept than Germany. Consequently, the questions to be answered include: What accounts for the conceptual differences? What accounts for the differences in timing (the point in time when the concept was introduced) and the speed of adoption? How can differences in implementation faithfulness be explained? Answering those questions clearly is a more analytical exercise.

26 I am grateful to Pascal Vennesson for raising awareness to this methodological problem and who argued that such categories might be useful analytical tools but should neither be seen as normative frameworks nor be confused with the actual adoption processes.

3.5 Explaining (Military) Organisational Change and Adoption

There are a number of theories that offer explanations of the phenomena of military diffusion and adoption. The various theoretical branches of international relations theory and organisational theory inspire these theories. Theo Farrell and Terry Terriff regarded military diffusion (labelled as emulation) as one of three possible pathways whereby military change can occur; the other two being innovation and what they call adaptation (adjusting existing military means and methods) (Farrell/Terriff 2002). Yet, military change can also mean sociological change. Military sociology research is concerned with subjects such as minorities within the armed forces or the reconciliation of work and family life (Moskos et al. 2000). And historians might yet have another agenda when they study military change (Lynn 2001).

Consequently, there are many recurring and competing views regarding the factors that contribute to military change or, more specifically, concept adoption. In the social sciences there are two dividing lines that structure the various arguments that seek to explain military change. The first dividing line is between arguments that see the main drivers for change or concept adoption at the international level and those that see them on the domestic level. Neorealist (Resende-Santos 2007; Waltz 1979) and constructivist accounts (Price/Tannenwald 1996) focus on drivers for military or strategic change stemming from the systemic, international level. Domestic accounts for strategic and military change focus rather, for example, on the institutional constraints of the political system (Avant 1994) the role of bureaucracies, the influence of junior officers on military innovation (Rosen 1991) and the influence of civilian policymakers on strategic decisions (Posen 1984). Opening the black box of domestic policy-making Posen's claim is that civilian leaders concerned about the balance of power engage in the change of military doctrine: "Statesmen will intervene in the doctrines of their military organisations as part of an overall pattern of balancing behaviour" (Posen 1984: 231).

The second dimension is concerned with arguments that focus on the assumption of rationality as a driver of change versus cultural and institutional arguments. One such rational driver for military change is the real and urgent existence of a threat. Another could be success in battle by one state that leads to the emulation of its practices or technologies by other states (Farrell/Terriff 2010: 10). Jack Snyder showed how military organisations would ideally assess the value of either offensive or defensive strategies on the basis of national aims, and assessment of technology and geography, and the military balance. He found, however, that in a number of cases the choice of the actual policy is "sometimes shaped by motives that conflict with the dictates of sound strategy (Snyder 1984: 24). Switching the focus from the strategic level to the organisational realm Snyder showed that military organisations are biased in their perceptions of threats and adequate responses by internal parochial interests and the need to make war planning "more manageable" and, thus, to "develop relatively simple but effective techniques for canning and organizing information about the problem, and for structuring and evaluating the available options" (Snyder: 27). While Snyder offered an alternative explanation to the rational calculus of strategic choice in terms of threat assessment and turned to the focus to processes within the military organisation his analytical framework nevertheless favoured a rationalist explanatory framework that focuses on bureaucratic functionalism.

Cultural arguments, on the other hand, also offer explanation for military change (Farrell 1998). Elizabeth Kier assumed that decisions within organisations and civilian decision-making "are framed by their perception of the world" (Kier 1997: 4). Thus, action is shaped rather by norms that serve as a guideline of what is considered an appropriate action and what not. Other works focus on organisational culture. In his book *Learning to Eat Soup with A Knife*, John Nagl (2005 [2002]) examined how the British and the U.S. armies learned when they were confronted with situations for which they were not prepared. Analysing the British military performance in Malaysia and the U.S. performance in Vietnam, he concluded that the British were much more successful in adapting their structures and processes because the British Army was a "learning institution" and the U.S.' was not (Nagl 2005 [2002]: x). Terry Terriff (2006) argued that, in the 1980s, the U.S. Marine Corps highlighted its amphibious warfare capabilities not so much out of strategic necessity, but because this would help differentiate the Marine Corps from the U.S. Army. James Corum studied the adaptation of technology (vehicles, communications and aviation) in the interwar period in the German and French militaries and found that Germany was more successful in developing and procuring new technologies due to an innovation-based military culture (Corum 1994). Others applied the concept of culture not to the organisational level of the military, but argue more generally that national security cultures influence the way a country approaches the question of armed conflict (Heuser 1998; Katzenstein 1996; Maull 1990).

Each of these literatures have developed their own views on what accounts for military change and argue convincingly that the respective drivers and levels of change do influence concept adoption. The concentration on only one explanation (systemic-rational, systemic-ideational, domestic-rational or domestic-cultural) masks, however, other likely factors that explain differences and variations. On the one hand, such a conceptual reduction is necessary in order to deal with complex social phenomena. Complexity reduction allows for the drawing of general conclusions. But still, explaining differences in concept adoption might benefit from insights that have been touched upon by the various strands in the literature. A *balanced approach* to explaining military change modifies the two dimensions that are found in the literature and develops more adequate notions of the level/location of the driving forces and the driving forces themselves. In this view variances in concept adoption stem from an interplay of international and domestic factors as well as from an interplay of considerations of rational effectiveness by the adopting organisations and considerations of institutionally-derived legitimacy. Potential driving forces for military change and concept adoption are existent on the international, national political and organisational level. Still, some factors might be more influential than others. The key to understanding what level poses what influence is more complex than just looking at the presence or absence of a factor, it is the way the focal organisation approaches every one of these factors. This view is similar to that of the "enacted environment" (Pfeffer/Salancik 1978). Here, it is assumed that an organisation determines its environment (it enacts it) and defines the factors it is conscious and responsive to. It is argued that to an extent (military) organisations deliberately respond to those factors in the process of NCW adoption.

The second important question is concerned with the rationale for action. What are the underlying drivers for change? The literature on military change broadly divides along rational arguments such as the existence of preferences, functional imperatives and strategic choice and institutional/cultural arguments where existing norms and ideas influence a course of action. This dichotomy is not limited to research on military change, as it can be found widely in economics and social science: Be it the logic of appropriateness versus the logic of consequence that drive human behaviour (March/

Olsen 1995, 2004); the differentiation between efficiency-increasing and legitimacy-increasing behaviour of social organisations (Meyer/Rowan 1991); the assumptions that there is such a thing as a homo economicus and that market exchanges are based on purely rational decisions – an assumption that has been criticised and qualified early on (Veblen 1899); or the materialist-idealist divide in international relations theory (Wendt 1999).

It is argued that this theoretical dichotomy is also not helpful in understanding differences in concept adoption as both perspectives put blind spots on factors that nevertheless contributed to the outcome that is to be explained.[27] Rational and institutional (and cultural) explanations are thus brought together. The chief argument is that the variance in concept adoption can be explained by the different composition of rational and institutional drivers in each case. Bringing rational efficiency arguments and social institutionalist arguments together is not uncommon. A study that benefited from such a perspective is Pamela Tolbert and Lynne Zucker's account of the diffusion of civil service reforms between 1880 and 1935. They found that that early adopters were rather driven by efficiency considerations, that is, improving their performance. Late adopters, on the other hand, were rather motivated (just) to *appear* as being efficient (Tolbert/Zucker 1983: 22): "[E]arly adoption of civil service by cities is related to internal organisational requirements, with city characteristics predicting adoption, while late adoption is related to institutional definition of legitimate structural form, so that city characteristics no longer predict the adoption decision."

A critic might object that a combination of both perspectives would result in a catch-all explanatory framework, unable to produce falsifiable outcomes. Yet, it is not argued that in order to explain NCW adoption every institutionalist or rational factor has to be included. Instead, it is posited that *difference in the composition* of influential factors (thus, those institutionalist or rational factors that were present and were responded to by the military organisation) can explain differences in military adoption.

The case studies will thus trace both rational and institutionalist motivations, incentives and constraints for action.

3.5.1 Efficiency and Effectiveness

Starting with the notions of efficiency and effectiveness – what does this notion encompass and how can the concept of efficiency be usefully adapted to help us understand military change? Merriam-Webster's collegiate dictionary (Merriam-Webster Inc. 1998) defines efficiency as a metric that compares the production with cost (as in energy, time and money). In economics, efficiency is often understood as productivity, that is, the ratio of output to input; the ratio of the good or service that is produced to what is required to produce the product.

Although military organisations are subject to economic concerns and often have to operate under a usually (annually) pre-defined budget, when it comes to operating in crises and war, military organisations do not follow purely economic considerations. Allison and Zelikow (Allison/Zelikow 1999: 149) describe the differences between commercial organisations and government organisations by stating that the latter, "are called into being by political processes; their goals – like their masters – are often diffuse. Government organizations are especially burdened by unique constraints; they cannot keep

27 I have argued in a similar fashion elsewhere (Wiesner 2010).

their profits; they have limited control over organization of production; they have limited control over their goals; they have external (as well as internal) rules governing their administrative procedures; and their outputs take a form that often defy easy evaluation of success or failure".

While military organisations do have to be sensitive to budget ceilings they do not have to make profit. Furthermore, military organisations have to pay attention to finite resources and how to use them in the most *efficient* way.[28] Even though financial concerns did play a decisive role for the introduction of NCW the influence of efficiency concerns is generally limited by specific requirements the military is subject to. These are, for example, the protection of own troops and civilians in combat or maintaining a sufficient number of well-trained troops and infrastructure in peacetime. As a result, economic efficiency is not quite as important for military organisations as it is for commercial organisations operating in a competitive market environment. Regarding the core task of a military organisation in a democratic state – to provide security against external threats – financial costs or the amount of material resources do not matter as much as in other fields, especially not if the survival of the nation is at stake. With regard to core military functions, it is instead more appropriate to think in terms of *effectiveness*. Effectiveness here means producing a decided, decisive or desired effect (Merriam-Webster Inc. 1998).

Turning the focus away from efficiency and to effectiveness, which is a more suitable concept to capture performance in military operations, how has this concept been approached so far and how useful could it be as an independent variable in this study? First, there is not much conceptual work available to assess overall military effectiveness in a comparative way. From a historical perspective, Allan Millett and Williamson Murray (Millett/Murray 1988) published a three-volume set that looked at military performances during World War I (volume 1), the interwar period (v. 2) and World War II (v. 3). Risa Brooks tried to uncover the sources of military effectiveness (Brooks/ Stanley 2007; Brooks 2003), whereas Soeters' et al edited volume (Soeters et al. 2010) on 'managing military organisations' devoted a section to "the monitoring of operational effectiveness". Stephen Rosen assessed the relationship between societal structures and military effectiveness. Starting from the question if soldiers from all societies from all cultures would fight the same way, he found that this was not the case and that indeed military effectiveness varied in conjunction with the dominant structures of society and the degree a military organisation would distance itself from these structures (Rosen 1995).

The possibly clearest disquisition on the subject of military effectiveness was issued by Allan Millett, Wiliamson Murray and Kenneth Watman (1986, also see Millett/ Murray 1988). They mainly employed both performance-oriented and functional perspectives on effectiveness.[29] Starting from a definition of effectiveness that incorporates

28 I am grateful for Pascal Vennesson for raising my awareness of this point.

29 Barton Cunningham pointed out that – depending on the focus – there are a number of alternative approaches on how to assess organisational effectiveness (Cunningham 1977). One approach, for example, focuses on the performance of the organisation is the rational goal perspective that assesses the organisation's ability to achieve its goals. Another approach focuses on the performance is the systems resource model, which is concerned with the organisation's ability to satisfy the needs of its components. Another focus is the performance of the organisation's human resources, and here the effectiveness of managerial processes and the organisational development are assessed. The bargaining model approaches effectiveness from the view of an orga-

a notion of efficiency, the authors argue that military effectiveness must be assessed at all levels on which military activity occurs (Millett et al. 1986: 37): "Military effectiveness is the process by which armed forces convert resources into fighting power. A fully effective military is one that derives maximum combat power from the resources physically and politically available. Effectiveness thus incorporates some notion of efficiency."

From their point of view, these resources cannot only be found at the tactical and operational, but also on the political and strategic levels. On the political level, a military organisation has to "secure the resources required to maintain, expand, and reconstitute itself" (Millett et al. 1986: 4). Here, the military organisation is in competition with organisations in other fields of public policy over resources (Millett et al. 1986: 38f.). Where strategic effectiveness is the translation of political goals into military strategy (Millett et al. 1986: 43), operational effectiveness is concerned with the analysis, planning, preparation and conduct of military campaigns (Millett et al. 1986: 50). Finally, tactical effectiveness refers to the specific techniques that are used in combat situations to achieve operational goals (Millett et al. 1986: 60). For each of these four levels, the authors developed a set of questions that would allow for an assessment of the respective effectiveness. For example, "[c]an military organizations assure themselves a regular share of the national budget sufficient to meet their major needs?" or "[t]o what degree are strategic goals and courses of action consistent with force size and structure?"

In this book, the understanding of effectiveness slightly departs from Millett et al. in that it treats effectiveness and efficiency as less combined. It is understood that military effectiveness describes the extent to which the military organisation is able to fulfil its core mission, whereas military efficiency describes the relationship between the resources necessary to achieve a desired military outcome. Nevertheless, the Millet et al. way to operationalise effectiveness had offered the staring point from which to develop a set of questions that could uncover the possible motivations of the Bundeswehr and the UK Armed Forces with regards to the introduction of NCW: For example, had NCW been considered as potentially decreasing budget pressure? Had NCW been considered as potentially reducing the need for manpower? Had NCW been considered as a means to achieve the strategic objectives?

These questions represent the operationalisation of the factor 'efficiency and effectiveness' as a driver of NCW adoption. The more questions that could be answered positively, the stronger the evidence that effectiveness (and, where applicable, efficiency) concerns have in fact been a driver for the adoption of NCW. Alternatively, the more questions that could be answered negatively, the less likely it is that concerns for military effectiveness played a roll for and during the adoption of NCW.

---

nisation as a resource-distributing system in which individual or groups pursue their own goals and assesses internal decision-making and bargaining processes. The structural approach understands organisational effectiveness as its ability to develop structures such as alliances, contracts, commitments and so on in order to survive. Finally, the functional approach traces how well the organisation's activities serve the needs of its client groups.

3.5.2 Institutional Legitimacy

It would be naive to assume that military organisations are mere rational optimisers of efficiency and effectiveness. Similarly, a military organisation's external environment is also never purely material, or, using the terminology by Scott and Meyer 'technical' (Scott/Meyer 1991). Like any other social organisation, military organisations act in and interact with their social environment. The mechanism by which the environment influences an organisation's course of action will be called *legitimation*. Maintaining and gaining social legitimacy is an important motor for organisational behaviour and for military organisations can thus be a motive for foreign concept adoption. Dana Eyre and Marc Suchman, for example, make the case that states may invest in state-of-the-art weapons that have a high symbolic significance although other, more mundane military equipment, would suit the specific strategic circumstances better (Eyre/Suchman 1996: 96).

In institutionalist theory, organisational action is constrained by regulations, rules and norms. Appropriate actions result in the organisation's behaviour being regarded as legitimate and thus provides for its survival (Suchman 1995: 574f.). In the organisational context, survival means the continuation of social support and the provision of (material) resources. Sometimes, institutional rules are perceived by the organisation as contradicting the pursuit of efficiency and the organisation might *decouple* itself from the institutional rules, making conformity to these rules only ceremonial and symbolic while informally structuring behaviour in order to maintain or achieve efficiency (Meyer/Rowan 1991: 57f.).

Recent works point out that organisations can seek legitimacy in order to gain reputation and "enjoy a favourable social evaluation" (prestige, see Deephouse/Suchman 2008: 66). Departing considerably from the logics of efficiency and effectiveness maximation, the quest for institutional legitimacy is the other, the second factor that moulds and shapes organisational change. Institutional legitimacy is the social acceptability and credibility that organisations require in order to "survive and thrive in their social environment". (Scott 2001: 237) Some would refute this view and claim that what we understand as efficiency and effectiveness itself is socially constructed (Powell 1991) Yet, assuming there is a difference between rational efficiency-seeking and an institutionally influenced quest for legitimacy does not run counter to every day experience. Organisations do have goals that are written down, military organisations have mission statements defining the very purpose of their existence and their actions are to a great extent directed towards achieving these goals and mission statements. Therefore, manifestations of efficiency and effectiveness can be recognised.

There is more than one conception of legitimacy. A rather strategic tradition (Pfeffer/Salancik 1978) treats legitimacy as a resource that organisations aim to gain (thereby even manipulating the perception of their own actions to make them look more appropriate). This is where conceptually at least the strive for legitimacy might be regarded as nothing else than a form of rational effectiveness-maximising organisational behaviour. Others see legitimacy as cultural pressures that are given and thus beyond the organisation's direct control (DiMaggio/Powell 1991 [1984]; Meyer/Rowan 1991). Others argue that depending on the situation organisations in turn can affect those norms, values, beliefs and definitions that constrain their behaviour (Allison/Zelikow 1999: 151). Mark Suchman proposed a combination of these views. For him, legitimacy is "a generalized perception or assumption that the actions of an entity are desirable, proper, or appropriate within some socially constructed system of norms, values, beliefs, and definitions" (Suchman 1995: 574).

A similar view is employed throughout the book: institutional legitimacy contains both aspects of desirability that is to some extent in the discretion of the organisation but also external expectations of appropriate actions.

The general existence of two broad motivations for organisational action is uncritical in most circumstances when effectiveness-maximizing behaviour is regarded as legitimate. The matter becomes interesting for social scientists, however, when effectiveness and legitimacy collide; when the organisation's ways to achieve certain aims are in conflict with external expectations. Faced with such a situation an organisation could either adjust its own practice, or enter into a strategic discourse with the aim of (re-)negotiating expectations or regulations. Another possibility for organisations to deal with undesirable but constraining norms is the decoupling of symbolic actions aimed at showing compliance to those norms from internal production modes (what the organisation actually does) (Boxenbaum/Jonsson 2008: 79).

In the literature, there are a number of different conceptualisations of what constitutes institutional environments (Scott/Meyer 1991: 109f.). Yet, the term 'organisational field' has become a widely accepted concept for theorising about institutional environments (Wooten/Hoffman 2008: 131). According to Scott, an organisational field is "a community of organizations that partakes of a common meaning system and whose participants interact more frequently and fatefully with one another than with actors outside the field" (Scott 2003b: 56). Focussing especially on the area of security policy and trying to create an organisational model for explaining government action Allison and Zelikow distinguished between primary (governmental organisations) and secondary (international world) environments (Allison/Zelikow 1999: 170).

Existing conceptualisations of the social environment implicitly or explicitly acknowledge focal organisations that are linked with the environment through exchanges, resource flows and through institutional processes. Yet again, these institutional processes are captured differently in the literature (Greenwood et al. 2008). Some see the institutional context as a cultural prescript (Meyer/Rowan 1991; Tolbert 1985), while others define it less value-ladden as regulative frameworks, rules and regulations (Héritier 2007; North 1990). Importantly, some works also address political power and agency within the environment (DiMaggio 1988; Oliver 1991).

How can the environment for military organisations best be understood? According to Scott and Meyer there are two fundamentally different kinds of environments: Technical environments "in which a product or service is produced (...) such that organizations are rewarded for effective or efficient control of their production system" and institutional environments that are characterised "by the elaboration of rules and requirements to which individual organizations must conform if they are to receive support and legitimacy" (Scott/Meyer 1991: 123). Usually, these environments coexist and overlap. Whether an organisation is more responsive to technical environments or institutional ones depends on the very nature of the organisation. According to Meyer and Rowan, organisations can be arranged along a continuum ranging from efficiency-seeking production organisations (usually commercial organisations) to institutionalised organisations such as schools or universities (Meyer/Rowan 1991: 55). Whereas technical environments are concerned with the effective and efficient control of production, institutional environments are composed of regulatory agencies of all kinds that set requirements to which an organisation must conform in order to receive external support and legitimacy.

Military organisations usually operate in an environment that is characterised by the mutual influence of technical and non-technical aspects. On the one hand, they operate in highly technical environments and are subject to considerations of effectiveness (i.e. manpower, capabilities) and effectiveness (i.e. mission success). Yet, they are also dependent on national (and international) support in order to legitimise their purpose, their current missions and rules of engagement. Problems are likely to occur when 'technical activities' stand in conflict with demands from the institutional environment (Meyer/ Rowan 1991: 56). In the case of NCW adoption the British and the German militaries were in fact faced with this challenge of different (and, in the case of Germany differing) environmental demands. Their responses to these external influences resulted in the very different adoption outcomes.

3.5.3 NCW Adoption Environments

What specific kinds of environments can be defined for the British and the German Armed Forces? Broadly speaking, the two military organisations act and interact on two distinct levels – the international and the national. On the international level, the organisational field is composed of professional military organisations. As both states furthermore are members of NATO and – to varying degrees – subject to U.S. military hegemony, it can be expected, that the UK and Germany are influenced by professional (Western) military norms and beliefs, by NATO rules and regulations, by NATO standardisations and capability goals and, more generally, by U.S. military policy. Internationally, the technical constraints are derived from a strategic threat assessment, operational contingency planning and actual operational planning. Furthermore, interoperability requirements for a coalition might result in technical constraints to military organisational decision making.

On the national level, the military organisation is dependent on domestic support and legitimacy. In democratic societies military behaviour has to reflect the values of the society and the preferences of the political level. This can include moral beliefs about the purpose, structures and procedures of military action. Nationally, technical, effectiveness-related aspects might include the core tasks set for the military by the government, which can range from territorial defence to stabilisation operations or expeditionary operation abroad and which the military organisation is expected to fulfil.

Additionally, spheres within the organisation itself can influence a process such as concept adoption or military change. Although strictly speaking not part of the organisational environment internal phenomena such as military inter-service struggles for resources (Grissom 2006: 910ff.) might influence the outcome of military concept adoption. While by definition these factors could be treated as external factors it becomes almost impossible to treat other internal constraints as environmental. One such constraint, for example, is path-dependency: existing practices, existing structures, organisational goals and sunk costs of previous decisions almost certainly influence the way a new concept is adopted by a military organisation (Thelen 1999, 2004). Also, organisational culture might influence how a foreign concept is perceived and hence, the way that it is adapted to fit existing beliefs (Farrell 1998, 2001). On the organisational level the awareness for costs, casualties, and operational and tactical effectiveness can be understood as technical constraints for organisational behaviour.

Table 5: Operationalisation of the Independent Variables Effectiveness/Efficiency and Institutional Legitimacy

Level/ Nature of Factor	Operationalisation	Organisational Response
International Level		
Effectiveness/ Efficiency:	NCW to increase military combinedness and interoperability with the US?	y/n
	NCW to increase military combinedness and interoperability with NATO allies?	y/n
	NCW as a mean to increase military advantage against adversary in current conflict?	y/n
	NCW to keep up with future potential competitors or adversaries acquiring NCW?	y/n
Institutional Legitimacy:	Presence of U.S. pressure/incentives to adopt NCW?	y/n
	Presence of NATO pressure/incentives to adopt NCW?	y/n
	Other international pressure/incentives to adopt NCW?	y/n
	NCW to increase international prestige?	y/n
National Level (Political)		
Effectiveness/ Efficiency:	NCW to decrease budget pressure?	y/n
	NCW to decrease manpower needs?	y/n
	NCW to achieve strategic objectives?	y/n
	NCW to strengthen national defence-industrial base?	y/n
Institutional Legitimacy:	NCW to fit budget priorities?	y/n
	NCW to maintain or change civil-military relations?	y/n
	Presence of defence industrial lobby pressure/incentives	y/n
National Level (Societal)		
Effectiveness/ Efficiency:	NCW to increase ability to protect citizen, territory and state function?	y/n
Institutional Legitimacy:	Presence of societal pressures/incentives to maintain or change the purpose of force?	y/n
Organisational Level		
Effectiveness/ Efficiency:	NCW perceived as a means to face likely enemies and current operations?	y/n
	NCW perceived as a means to provide better jointness, higher mobility and flexibility?	y/n
Institutional Legitimacy:	NCW to create impression of reform?	y/n
	NCW a continuation of previous reform attempts?	y/n
	NCW consistent with the existing military culture?	y/n
	NCW consistent with the military culture of the single services?	y/n

It has been briefly touched upon above that it is not only the absence or presence of certain pressures or incentives from the institutional environment that influence organisational action. Pfeffer and Salancik argued that organisations could not survive by "responding completely to every environmental demand" (Pfeffer/Salancik 1978: 43). Instead, organisations might choose which external stimuli they are going to respond to.[30] By this, organisations *enact* their environment (Pfeffer/Salancik 1978: 63). Thus, simply making assumptions about the absence or presence of environmental factors might not be sufficient to understand why military organisations behave differently, or, as in the case of NCW diffusion, differ in the adoption outcomes. It will be furthermore important to look for the organisational response to environmental constraints or incentives.

To summarise, as military organisations operate in environments that are defined both by technical and institutional logics it would be misleading to look for only either of these dimensions. Instead, variance in the outcome (adoption patterns and pace), it is argued, can best be explained by differences in the combination of effectiveness and efficiency factors and factors of institutional legitimation. For the purpose of this study the environment has been operationalised with the help of standardised questions that are presented in table 5.

To refine the argument further, it is argued that differences in the outcome of NCW adoption in the UK and Germany can be understood as the result of different combinations of efficiency/effectiveness-guided or legitimacy-guided behaviour of actors involved in the adoption process. Varying degrees of efficiency/effectiveness-oriented motivations and legitimacy-oriented motivations, therefore, are likely to cause differences in the adoption process.

3.6 Conclusion

This chapter introduced the main theoretical aspects of diffusion and military change. Building upon the theoretical framework in the following two chapters the adoption of NCW in German and the UK is traced. Each of the following chapters contains a 'story line' about how NCW found its way into the British and German armed forces. That story line is structured along the respective adoption phases: knowledge acquirement, utilisation, and implementation. Attention has been paid to singling out those factors – efficiency/effectiveness-related or institutional legitimacy – likely to explain variance in the adoption NCW outcomes in the two states between 2001 and 2010. The book first turns to the UK, which – compared to Germany – presents a case of pragmatic and fast NCW concept adoption.

30 Pfeffer and Salancik (Pfeffer/Salancik 2003: 44) presented a list of conditions that would affect the extent to which an organisation would comply with external control attempts.

4 Network-Enabled Capabilities in the UK

The British variant of Network-Centric Warfare was called Network-Enabled Capabilities (NEC). The NEC concept emerged in 2002 and was officially mentioned for the first time in the New Chapter to the Strategic Defence Review later that year (British Ministry of Defence 2002). Implementation of NEC started almost in parallel with the drafting of the British conception in 2003. Since then, NEC had been dutifully introduced into the British Armed Forces. In 2011 the Ministry of Defence suggested to end the NEC programme as its main goals had been achieve. The development of network-enabled capabilities was then considered "daily business" in the frame of a broader information superiority programme. Yet despite the apparent success of NEC there was also a significant debate concerning NEC and its application (Potts 2003; Storr 2009).

NEC differed considerably from the original U.S. NCW concept (Meiter 2006). One difference concerned the purpose and ambition of NEC, which in the British view was more about *enabling* (better) military functions than trying to *determine* the way the military operated. Apart from a much humbler view on what could and could be achieved with NEC, the British concept, moreover, put more weight on the human dimension of war than the U.S. counterpart. For the British, NEC was not so much a technological, but also a social concept.

This chapter is concerned with the NCW adoption process in the UK between 2001 and 2010. It aims at re-creating the processes by which the British military became aware of the concept, how the concept was utilised and how it was implemented. The chapter also analyses the unique British features of NEC and explores how it differs from the U.S. concept.

4.1 The NEC Adoption Process in the UK

The introduction of NEC into the British Armed Forces was not a revolution nor was it particularly disruptive. Rather, it was the continuation of previous modernisation attempts and the "latest in a series of initiatives aimed at digitizing the UK Forces" (Defence Science and Technology Laboratory 2003). Earlier digitisation initiatives comprised of the "Digitization of the Battlespace", the "Joint Command Systems Initiative" and the "Joint Battlespace Digitization", all developed during the 1990s (Ferbrache 2003). The Joint Command Systems Initiative started in 1995 and was intended to provide integrated command, control, communications and intelligence using a joint approach. Besides an organisational dimension of the programme, the main aim was to establish world-wide deployable communications and information systems (British Ministry of Defence 1996). The UK MoD's Joint Battlespace Digitisation (JBD) initiative had been announced in the 1998 Security and Defence Review. While NEC shared many of the features of these earlier initiatives, it is unprecedented in its comprehensive, defence-wide and integrated character.

The British conception of NEC emerged between 2002 and 2003. The importance of the network-centric capabilities was mentioned in the New Chapter to the Strategic Defence Review 2002 (British Ministry of Defence 2002). In this document the British idea of net-centric forces was still shallow and mainly a repetition of the U.S. NCW conceptual tenets and promises. On the other hand, however, the issue was treated in length compared to other aspects covered in the document – an indication of the interest NCW attracted. The New Chapter to the Strategic Defence Review is seen by many as the hour of birth for the British NEC initiative (Baxter 2005; Ferbrache 2003; Fulton 2003), as it established an "overarching priority" on systems integration to achieve precision engagement (Williams/Foley 2003: 168). New types of C2 equipment singled out in the New Chapter of the Strategic Defence Review finally led to a re-focussed equipment programme in 2003 that enabled significant new investment in NEC. While this first mentioning of the importance of network-centrism for military operations in the New Chapter did not quite contrast the British approach from the original U.S. concept, the British NEC concept was hammered out in the months following the release of the New Chapter.

The 2003 Defence White Paper, *Delivering Security in a Changing World*, – the next top-level strategic document that was issued one year after the release of the New Chapter – finally and officially introduced NEC and stressed the importance of this concept for future UK military operations (British Ministry of Defence 2003a). NEC was similarly prominent in the 2003 UK Joint High Level Operational Concept (Joint Doctrine and Concepts Centre 2003) and, in due course, the individual services included NEC into their future operational concepts (British Army 2008; Royal Air Force 2006; Royal Navy 2004). Also from 2003 onwards, a considerable effort was made by the MoD to bring NEC to the theatres of Iraq and Afghanistan. All of the people interviewed for this book stressed that the overarching priority of the NEC initiative (as has been the case with earlier digitisation initiatives) was to quickly equip deployed soldiers with the best means to conduct their mission.

The implementation of NEC started in 2003. In fact, procurement projects, such as the Bowman communications system, that had been initiated much earlier during the digitisation period in the 1990s were re-labelled as NEC projects. From the perspective of the MoD, NEC implementation appeared to be a success (British Ministry of Defence 2009).

4.1.1 Knowledge Acquirement

In the UK, the adoption of NEC presents itself as a case of continuing attempts to integrate tri-service command-and-control systems.[31] This focus on the importance of command-and-control systems is a result of the Gulf War experiences in 1991 when the UK was part of the U.S.-led coalition and thus subject to interoperability requirements[32]. During the 1990s MoD documents therefore stressed the importance of information and information technology in warfare (British Ministry of Defence 1996) and the need for adequate C4ISR capabilities (British Ministry of Defence 1998). The First Gulf War thus presented the vantage point after which the capability focus was shifted to command, control and communications (C3) systems. These were later also at the heart of the NEC initiative.

Apart from the technological dimension, another important area of NEC is doctrine. During the 1990s, the MoD started to foster greater service jointness – not only strategically in the Ministry, but also at the tactical and operational levels. Yet, in 1996, when the Permanent Joint Headquarters (PJHQ) and a joint rapid deployment force had been established (Connaughton 2000), it became clear quite quickly that an adequate, joint

31 This section greatly benefitted from interviews (London 10 November 2009, 18 November 2009).

32 See the various articles in the special issue of the journal *Military Technology* (1991: 4).

command-and-control structure was missing to command operations abroad. The individual services did run their own systems inside the PJHQ and were often unable and sometimes unwilling to share information properly between each other. As a consequence, the MoD issued the Joint Command System Initiative in 1995, aiming at setting up a single command, control and intelligence (C2I) system to connect the MoD, the PJHQ, the individual services and deployed headquarters (British Ministry of Defence 1996). This system was called JOCS – the Joint Operational Command System. Implementation of the JOCS began in 1999 and the system was used during the invasion of Iraq in 2003. JOCS relied on commercial off-the-shelf products (COTS) purchased as urgent operational requirements (UOR).

At the same time, the UK experienced a shift in its operational challenges. The conflicts that followed the First Gulf War in 1991 were different from Cold War scenarios with clear front lines and a clear distinction between civilians and combatants. The Bosnia experience showed that those civil or secession wars would require less air force or naval engagement, but place the operational burden first and foremost on the Army. As a result, the Army strove to become more flexible and modular in operations. A flexible and modular C2 structure was thus in need. The Army set up its own C2 modernisation project called "Digitization Battlespace (Land)" (DBL), mainly based on the – yet to be implemented – Bowman and Falcon communications system projects. The Army was inspired by the U.S. Army modernisation project 'Force XXI', which aimed at setting up a digitised brigade.[33] To support its C2 modernisation project, the Army established the C2 Development Centre in Warminster in 1997. Ultimately, however, the DBL was discontinued and the JBD initiative, which aimed at integrating Land, Sea and Air C2 components replaced both JOCS and DBL. JBD could well be identified as the predecessor of NEC.

4.1.1.1 The Joint Battlespace Digitisation (JBD) Initiative

Governance of the JBD project rested with the Deputy Director of Equipment Capability for Command and Control Information Infrastructure (DDEC CCII). His team in the Ministry consisted of about 20 people. Their tasks did not only comprise of the management of the JBD initiative, but also of the conceptual development of the UK's approach to command and control. To this end, the DDEC CCII worked together with the Defence Scientific Advisory Council (DSAC), an independent, non-departmental body that advises the Cabinet in the areas of science, engineering, technology and analysis. In 1998 and 1999, the DSAC published two reports on the C2 aspects of digitisation and information management.

By the end of 1999, the JBD initiative was in serious trouble, as it had not delivered the expected output. This was mainly for three reasons. First, compared to other equipment projects, C2 was generally different in nature. While a new ship, a new aircraft or tank is a quite tangible subject for negotiations about budget shares for new procurement projects, the JBD initiative comprised of a number of smaller projects in the area of command and control systems and therefore did not possess the same bargaining power as bigger projects. If a project comprises of a number of systems, in the course of

33 Interview, Warminster, 16 June 2009.

negotiation some of them might be cancelled or downsized more easily than this would be even possible in the case of one major weapon project (generally on this problem see Kincaid 2008: 278; Mahnken 2008: 179).

The second problem of the JBD initiative was the different internal structuring of the Defence Procurement Agency (DPA, the supplier) and the Equipment Capability Customer (ECC, the sponsor of acquisition) in the MoD. Whereas the former was structured around major weapon projects, the latter was structured along the lines of military capabilities. This resulted in overlapping competencies and different motivations, as the reward systems for the directors in the various DPA leaders were different from those of the Directors of Equipment Capability in the MoD. Third, the JBD initiative focussed almost entirely on equipment and neglected non-equipment issues such as training or information management (Baxter 2005). The acuteness of finding a solution for the UK's insufficient C2 system jointness became apparent when the air campaign in Kosovo revealed serious flaws in the existing communication and intelligence systems.

As a result of the difficulties, the JBD initiative faced Vice Chief of the Defence Staff (who, together with the 2nd Parliamentary Undersecretary 'runs' the MoD Equipment Capability Customer), Admiral Peter Abbott, set up a JBD steering group in 1999 to overcome diverging interests of the single services in this matter. This group comprised of high-level military officers such as the Deputy Chief of the Defence Staff (Equipment Capability), the Chief of Joint Operations (who headed the PJHQ) and the Assistant Chiefs of Staffs for the Army, Navy and Air Force and met on a regular basis. Later, in 2001, when the JBD initiative was still in trouble, the efforts were integrated into the new Command and Battle Management (CBM) programme within CCII. Here, the importance of non-equipment dimensions such as training and information management received more attention.

4.1.1.2 First Awareness of NCW

It was around that time – in 1999 – that the new idea of NCW that just had been published by the U.S. DoD CCRP entered the British MoD. Quite naturally, JBD team became interested in the new concept. Frederick Stein, the co-author of the 1999 book *Network-Centric Warfare*, had personal ties with the Deputy Director of Equipment Capability, as both had served in Serbia during the Kosovo War. In the year 2000, the wider British defence community was provided with information on NCW when the *RUSI* journal published an article on the value of NCW for coalition operations (Gormley; Hart 2000).

NCW was perceived rather critically at first by the MoD staff. For example, the UK Joint Vision that was released in June 2001 and that offered a long-term vision for the British Armed Forces, did not explicitly mention NCW theory (Joint Doctrine and Concepts Centre 2001). Nevertheless, it referred to the importance of achieving a Joint Operational Picture and envisaged that "[d]igitisation will enable dispersed military units to operate as part of a virtual and dynamically reconfigurable network to improve the tempo and lethality of operations" (Joint Doctrine and Concepts Centre 2001: point 212).

The reasons for this rather lukewarm reception of NCW were twofold. First NCW was an original U.S. concept – a circumstance that the British with their own distinct military culture can be initially suspicious of.[34] NCW, furthermore, was considered a concept that put technology and a physical network at the centre of warfare and not – as the British perception of war would suggest – the human (see for an early discussion Blackham 2000b). Yet, despite the critical awareness within the MoD, the JBD team started to exchange ideas on networking with the U.S. Army and the Pentagon.

4.1.1.3 Active Knowledge Acquirement

In 2001, the U.S. Department of Defense set up the Office of Force Transformation (OFT). The OFT shaped the NCW vision further and, ultimately, NCW became the motor of the U.S. military transformation. Members of the JBD team were now regularly invited to Washington to give presentations on the British JBD. Similarly, the OFT director, Arthur Cebrowski, and his deputy, John Garstka, held NCW briefings at the British Ministry of Defence (Farrell/Bird 2010: 40). The OFT was interested in learning more about the C2 projects and concepts of its allies. For the MoD, these visits to Washington, in turn, presented an opportunity to learn about the conceptual underpinning of the American NCW concept. The UK embassy in Washington was similarly active in promoting the exchange of ideas concerning NCW and JBD at that time. The embassy not only organised exchanges between the MoD and the Pentagon, but also invited members of the JBD team to talk to the single-service constituencies in Washington, thereby allowing for a more balanced view on the U.S. services' attitude towards NCW at that time. The views on NCW were brought back to the UK by the JBD team. Finally, in November 2001, the British Defence Chief of Staff first mentioned the potential importance of NCW for the British Armed Forces in an official statement.[35]

The JBD team might have been the main gateway of information on NCW, but it has by far not been the only source. Almost 400 military and civilian personnel were located across the U.S., working as liaison or foreign exchange officers in U.S. joint or individual service headquarters. Although knowledge acquirement on NCW was centralised at the MoD, those working in respective C2 positions in U.S. headquarters witnessed U.S. digitisation as well and brought U.S. ideas back to the UK.[36] Another important forum of the UK-U.S. exchange of ideas on C2 at the working level was the Coalition Agents Experiment (CoAX). CoAx was an international collaborative research programme that ran from February 2000 to October 2002 and was comprised of ministerial and industrial partners from the UK, the U.S. and Australia (Allsopp et al. 2003). The experiment carried out a series of technology demonstrations on command and control in a coalition scenario, with a special focus on shared situational awareness and understanding – two important elements of the NCW value chain.

During the years 1999 and 2000, JBD and the new NCW concept had mainly been discussed as means to increase inter-service jointness in the UK. Then 11 September happened. The U.S.-UK campaign in Afghanistan made a strong case for combined interoperability with the U.S. forces. Suddenly, the challenges of a coalition operation increased the pressure on JBD (which by then had become part of the CBM). Yet despite the two years of work on JBD, the MoD "still had no network concept in 2001."[37]

34 Interview, London, 11 November 2009.
35 See http://www.iwar.org.uk/rma/resources/uk-mod/nec.htm (accessed 24 November 2009).
36 Interview, Warminster, 16 June 2009.
37 Interview, London, 11 November 2009.

The new chapter to the SDR, which was drafted after the terror attacks of 11 September 2001, presented a good opportunity to overcome the problem of a missing network concept and the dead-end in which JBD and its successor CBM appeared to be stuck. It was in these months before the issuing of the New Chapter that the British NEC approach had been hammered out. Responsibility for drafting the first NEC concept lay with the Equipment Capability Manager (Information Superiority) Maj. General Rob Fulton (who, in 2003, was promoted to Lieutenant General, on the appointment as Deputy Chief of the Defence Staff for Equipment Capability) and Brigadier Philip Pratley (DEC (CCII)). They were central intellectual figures in the development of NEC in the UK.

4.1.1.4 Conclusions

What conclusions can we draw from the process of knowledge acquirement? Which assumptions on the adoption pace can we make and which factors can we carve out that offer explanations for the findings?

The knowledge acquirement started early and the MoD Central Staff was well aware of NCW right from the beginning. From its very formulation in the late 1990s, NCW had been discussed inside the MoD. This was due to several reasons. First, within the MoD, there was already a strong awareness on the importance of information technology for military operations since the First Gulf War and the demonstration of superior U.S. technology (British Ministry of Defence 1998). This factor clearly bears a notion of effectiveness, but also of institutional legitimacy due to its path-dependent nature. Second, the MoD sought to further increase service jointness in order to be better prepared for the new kinds of military operations in the changed security context after the Cold War. A number of initiatives, for *jointness* in operations and acquisition, had already been launched and institutionally framed the introduction of NEC (Taylor 2009). As a result, the MoD was keen to find concepts and technologies that would allow for an integration of the individual service C2 structures to increase military effectiveness. Both of these factors are supported by the fact that acquisition structures for joint C2 were already set up within the MoD Equipment Capability Costumer when NCW entered the stage. Hence, in the British case, knowledge acquirement of NCW was a central activity in the heart of the MoD. During the interviews, no evidence could be found for a particularly steered knowledge acquirement activity of one of the individual services. The MoD Central Staff took the initiative early on. In consequence, there was no need for the single services to act as entrepreneurs in NCW knowledge acquisition.

A third factor for the short time period between the dissemination of the U.S. concept and the awareness of the British MoD presumably was the general awareness for new concepts from the U.S. – in the case of NCW, this awareness was even higher, as the concept received considerable attention in U.S. military circles and continued the prominent debate concerning a Revolution in Military Affairs. It can be argued, therefore, that NCW also received attention from the MoD because it has been disseminated in the U.S. and has been widely discussed there. This clearly would be a case of institutional legitimacy. It also contains, however, a notion of effectiveness, as being up-to-date with developments of the closest ally is important to secure influence on the ally's military decision-making processes.

Table 6a: The Timing of NEC Knowledge Acquirement in the UK

Timing of Knowledge Acquirement	Early
Explanatory Factor	Nature of the Factor
Decade-Long Awareness for Information Technology	Effectiveness and Institutional Legitimacy
Strong Joint Principles	Institutional Legitimacy
Aim of Joint C2 Structures	Effectiveness
Dissemination by the U.S.	Effectiveness and Institutional Legitimacy

No evidence could be found for an active and directed promotion of NCW by the U.S., whether directly or through the channels of NATO or, ABCA – an institutional setting between the armies of the United States, Britain and Canada, Australia and, to some extent, New Zealand. Quite the contrary, it was the British MoD who was eager to obtain information on NCW and discussed it in light of its own JBD initiative. Also, defence industry interests supposedly did not contribute to the early knowledge about NCW in the MoD.

Table 6b: The Pace of Knowledge Acquirement of NEC in the UK

Pace of Knowledge Acquirement	Moderate
Explanatory Factor	Nature of the Factor
Critical of Technology-Oriented Military Operational Concepts	Organisational Culture
JBD Project still Underway	Effectiveness and Institutional Legitimacy
NCW not yet Taken off Defence-Wide within U.S. Armed Forces	Institutional Legitimacy

Between 1999 and 2000, NCW had not been officially endorsed by the MoD and further knowledge acquirement proceeded only moderately, mainly through the JBD team. This was due to at least three reasons. First, the British military has a natural tendency to be critical of U.S. military operational concepts, which often are regarded as too technology-focused. This specific feature of British military organisational culture resulted in a critical perception of NCW. Second, the British JBD project was still running and efforts went into pushing JBD further instead of migrating new and foreign ideas. This factor bears both notions of efficiency and institutional legitimacy. Third, NCW was not yet more than an idea – only in 2001 did it become a motor for U.S. force transformation. Keeping still and awaiting actions of the senior military partner could be a sign of maintaining institutional legitimacy.

Finally, it seems that personal contacts played a major role in facilitating the UK-U.S. exchange of ideas on NCW. Examples include the personal acquaintanceship between one of the original NCW advocates in the U.S. and the then-Deputy Director of Equipment Capability in the MoD. There were also friendship ties between the JBD and staff members at the British embassy in Washington. With a view towards his professional career, one former military officer, concluded: "It all depends on whom you went to Staff College with".[38] The next section will look at the second phase in the NCW adaptation process and elaborate on the utilisation of NCW and the conceptualisation of NEC.

4.1.2 Utilising NCW as NEC

NEC was established as an acronym only days before the publication of the Strategic Defence Review: The New Chapter in July 2002. The idea of NCW gained prominence inside the MoD after 11 September 2001 and was considered as an alternative to the existing (yet deadlocked) JBD initiative when operational pressures grew due to the deployment of British troops to Afghanistan. Major responsibility for writing the sections on the British approach of NCW for the 2002 New Chapter lay with the Capability Manager (IS) General Fulton and the DEC (CCII) Brigadier Pratley.

4.1.2.1 Introducing NEC to the British Defence Community

Fulton and Pratley reportedly based their initial thoughts about NEC on the American NCW concept, but adjusted it to resemble British military thinking. As much as they were convinced of the 'network' idea as such, they struggled to warm to the notion of 'centric'. According to British military thinking, the human is the centre of all military action and not some technology or network. Another problem was the 'warfare' concept. For an Army that considered itself a "force for good" the notion of warfare had too much of an aggressive connotation. "Warfare" thus became "capability". The term capability was quite often used during the 1990s when, in the light of the strategic shift, the procurement focus changed from platforms to capabilities (see on capability-based planning Davis 1993). As a result, Fulton and Pratley soon ended up with the abbreviation *N[?]C*. They chose the *E* for *enabled* – a word that was quite popular in the MoD sphere, as it had been commonly used in various McKinsey reports on defence procurement in the early 2000s. In the end, their contribution to the 2002 New Chapter introduced NEC to the public and roughly sketched out the new approach.

A short anecdote may be helpful here to gain some insight on the development of NEC in Britain. A close observer of NEC history in the UK might object and argue that the 2002 New Chapter does not actually speak of NEC but of NCC – Network-Centric Capability. Yet, as frustrating as it was for Fulton, Pratley and the CCII team – the term Network-*Centric* Capability was a typo that has been written into the New Chapter of July 2002 by a proof reader who was familiar with the original NCW theory and thought Fulton and Pratley had just made a mistake. Later on, *NCC* never appeared anywhere else.

Fulton and Pratley were keen to emphasise that NEC differed in scope and focus from NCW. Yet, in 2002, all types of ideas concerning the nature of NEC were out in the open. Despite the scepticism of the CCII team about equating NEC with NCW, the then-Deputy Chief of Defence Staff (Equipment Capability) Air Marshal Stirrup, for example, explained in a meeting with the House of Commons (HoC) Defence Committee that there was essentially no difference between NCW and NEC and he showed himself confident that, despite some technological delays, the UK had an advantage

38 Interview, London, 11 November 2009.

vis-à-vis the U.S., as its forces were smaller and more joint and could more easily change concepts and doctrines (Lake 2003).

Briefly after issuing the 2002 New Chapter, Fulton issued a plan for the delivery of NEC, with this plan being endorsed by the Joint Capabilities Board (JCB) in October 2002. In the conceptual work that followed the CM(IS) high-level mission statement for NEC in 2002, the original U.S. NCW value chain was used as a starting point to derive so-called 'enablers of network centricity' (Alston 2003). These enablers were then summarised in the 2003 NEC Outline Concept as the "Nine NEC themes" (i.e. agile mission groups, synchronised effects, etc.).

A preliminary NEC model, also known as the London Underground Map, was developed, graphically illustrating to the JCB the envisaged information network that linked the key components of the equipment capability. At the end of 2002 an NEC flyer introducing the concept and setting out the vision, benefits and core themes to the wider defence community was also issued. Research on social and organisational aspects of NEC, furthermore, was initiated and conducted by the MoD Defence Science and Technology Laboratory, QinetiQ[39] and with academic support from civilian universities (Holt 2003).

4.1.2.2 Drafting the NEC Concept

The conceptual development of NEC can be traced by evaluating four open-source MoD publications. First, in May 2003, the MoD published the "NEC Outline Concept" which introduced NEC, set out its origin and purposes and contrasted it with the U.S. Network-Centric Warfare concept. This Outline Concept was informed by the work of the NEC Delivery Team, which was led by the Defence Science and Technology Laboratory (DSTL) with support from QinetiQ and the defence industry. As a result, the Outline Concept was comparatively technical and less comprehensive than later NEC concept papers. Second, MoD thinking in 2003 is also captured in a special issue of the Journal of Defence Science from September 2003, which hosted authors from the MoD, QinetiQ and the DSTL and touched upon a number of dimensions of NEC conceptualisation and implementation. Third, the 'NEC Handbook' was issued in 2005 (British Ministry of Defence 2005). This handbook formed the authoritative basis for the British NEC concept. Fourth, in January 2009, the handbook was partly reviewed and the results of this review were published by the newly established NEC Programme Office in the form of an edited MoD pamphlet (British Ministry of Defence 2009). In it, some minor conceptual changes on NEC are reflected (i.e. adding a fourth dimension to the NEC dimension-model, emphasising non-technical aspects of NEC).[40]

In the following section, the 2003, 2005 and 2009 documents outlining the British NEC are analysed and compared with the original U.S. conception of NCW as laid down in the U.S. DoD NCW Report to Congress of July 2001 and the OFT/OASD-NII

39 The founding of QinetiQ was the result of the Smart Procurement Initiative in the late 1990s. The Defence Evaluation and Research Agency was privatised and many of the 12,000 scientific personnel were outsourced into the QinetiQ company that formed in 2001.

40 For a brief overview of NEC, consult Iain Standen, Deputy Head of the Network Enabled Capability Programme Office (NEC PO), who lays out the basic tenets of the NEC concept and the medium- and long-term implications in a RUSI DS article (Standen 2010)

Network-Centric Operations Conceptual Framework Version 2.0. (Garstka/Alberts 2004; U.S. Department of Defense 2001).

4.1.2.3 NEC and NCW Compared

The British NEC concept differs from the original NCW (later NCO) concept in a number of ways. It is less ambitious, pursued a different aim and applied a more comprehensive view towards the application of NEC; furthermore, it did not introduce the NCW domain model, but instead developed an NEC Dimensions model. Finally, the NEC concept resembled structural features of the MoD's view on military capabilities and was thus, from the outset, focused on the implementation of the concept.

First, the UK took a more cautious and less ambitious approach to networking. NEC was not made the heart of a British military transformation (there was no outspoken transformation project) nor was it seen as a new doctrine. Rather, it was envisaged as an enabler – not only for future capabilities, but especially for current operations (Meiter 2006: 191). A quote from the NEC Outline Concept illustrates this: "NEC shares the tenets of NCW, but is more limited in scope in that it is not a doctrine or vision. Nor does it seek to place the network at the centre of capability in the doctrinal way that the term NCW implies" (Defence Science and Technology Laboratory 2003). On the operational-tactical level, the notion of superiority, which was so prominent in the U.S. concept of NCW, was largely missing in the NEC concept, which, in the 2005 Handbook, was defined as "offering decisive advantage through the timely provision and exploitation of information and intelligence to enable effective decision-making and agile actions" (British Ministry of Defence 2005). As a result, NEC appears less ambitious than NCW.

Second, NEC had a different aim than NCW. Whereas NCW ultimately aimed at achieving combat superiority, NEC aimed at "the realisation of emerging UK Doctrine including Effects Based Operations (EBO), Battlespace Exploitation and Knowledge Superiority" (Defence Science and Technology Laboratory 2003). In the 2005 Handbook, NEC was envisaged as a means to support the implementation of Effects-Based Operations (EBO), which in 2005 had been a quite popular concept for military engagement (Farrell 2008). In the 2009 edition of the handbook, the enthusiasm for EBO had lessened and NEC was rather seen as an enabler for Joint Action than for EBO.

Third, NEC was more comprehensive than NCW. From the outset, NEC was envisaged to support military defence-wide functions and combat engagement. Even though the 2003 conception of NEC also mentioned some tactical examples of NEC such as the sensor-to-shooter link, NEC was also concerned with a wider applicability of the network: "The term NEC also implies the wider utility that networking has across the spectrum of operations and not solely to war fighting" (Defence Science and Technology Laboratory 2003). The 2005 Handbook outlines how all three levels of command – the strategic, operational and tactical level – might benefit from NEC. This is quite different from the original U.S. NCW concept, whose focus was rather tactical and not at all comprehensive (meaning the inclusion of non-military agencies into planning processes). In contrast, based on the ability to share information throughout defence, the MoD saw itself being enabled at the strategic level to conduct more coherent planning collaboration with other agencies and coalition partners. The operational level would benefit from the robust networking between the Permanent Joint Headquarters, the force headquarters and the higher HQ of NATO, coalition partners and the UK Front Line Commands. Here, the multilateral perspective on international security shined through –

something that was also missing in the original U.S. conception. At the tactical level, NEC was envisaged to improve the presentation of information to commanders, contribute to the understanding of the command intent and integrate sensors, decision-makers and effectors.

The NEC Handbook furthermore stressed the vertical links between the three command levels without, however, engaging in a debate concerning the principle of mission command – a military leadership principle, which was introduced into the British Armed forces in the 1980s (Shamir 2008). Quite the contrary, NEC was simply defined as supporting a manoeuverist approach, especially at the tactical level, by the network-wide expression of the command intent.

Fourth, in the 2005 Handbook, no explicit reference was made to the original NCW concept, the NCW value chain or the domain model – yet the idea of the value chain was nonetheless prevalent. One paragraph, for example, reads command and force elements will be progressively integrated into an interoperable information and intelligence infrastructure, which will form the essential 'kernel' of NEC. This will ensure robust "command and control, enable rapid dissemination of command intent, and generate Shared Situational Awareness. In turn, we will achieve winning tempo through Decision Superiority, which will permit the highly precise and agile employment of capabilities to achieve desired effects".

An 'NEC Benefit Chain' was introduced that was a loose reproduction of the NCW value chain, but with the difference that it largely omitted the notion of superiority. Instead it rather focused endogenously on *better* networks, information sharing and so on. Again, this is a strong indicator for a much humbler, less ambitious approach to NEC. The benefit chain, furthermore, did not end with more mission effectiveness, but with better effects, thereby linking NEC with the British effects-based approach.

Fifth, the NEC handbook did not pick up the idea of the 'domains of warfare' of the original U.S. conception. Instead, it introduced three overlapping and mutually dependent dimensions of NEC – the network, the information and the people. Four years later, in the modified NEC conception, these three dimensions were embedded in a fourth dimension to which they all contribute – joint action (James/Meyer 2009). Joint Action is defined as the end-effect, whereas the three NEC Dimensions contribute to its delivery (Coward 2009).

The first domain – the network – was defined as constituting the heart of NEC. Yet, contrary to the original U.S. NCW concept, the network in NEC would provide a pan-defence capability. In providing the network, the MoD chose to follow the U.S. approach by developing a Global Information Infrastructure and Defence Information Infrastructure (the latter referring to the DII programme that aims at modernising the whole information equipment of the MoD and the armed forces). It is quite important to note that, in the NEC concept, the network did not merely consist of C2 systems and information technology, but included applications, processes and people. Whereas the original U.S. NCW concept assumed that information superiority almost automatically leads to decision superiority, the NEC concept paid attention to the underlying processes (the information dimension). Accordingly, emphasis was given to information management and decision support tools – the second domain – to reduce information overload and to exploit the advantages the network could offer. The concern for information management already existed before the conceptualisation of NEC. The Strategic Defence Review issued in 1998 already warned that "the sheer volume of information will bring its own problems. It will only be useful if we can get the information to the right place at the right time and relate it to the wider military and political context" (British Ministry of Defence 1998).

The 2005 NEC Handbook similarly emphasised the role of adequate training (the people dimension); it read: "Whilst the Equipment Programme will deliver the hardware and software, the real benefits will only be realised over time with the sustained training and education of all Defence personnel" (p. 9). This focus on having the 'right personnel' to exploit information technology was also not new. During the days of the JBD, a strong emphasis on the people dimension was prevalent (Blackham 2000). The operational reality in Iraq (2003–2009) and Afghanistan (2001–), however, increased the focus on the human dimension of NEC. Operational reports raised concerns about insufficient NEC training (Griffin/Whetham 2007) and therefore training was given priority in the 2009 NEC Handbook.

Sixth, the 2005 handbook went on to introduce a rough implementation scheme for NEC envisioning three states of NEC maturity. The *initial state* was characterised by minor organisational changes and equipment enhancements, but would interconnect existing systems. The *transitional state* would see major organisational changes and the integration of systems to improve shared situational awareness. The *mature state* would be characterised by the "optimal exploitation of information, delivered through developed doctrine, organisations, process and equipment, together with personnel appropriately selected, educated and trained" (British Ministry of Defence 2005: 10). In 2006, time targets were developed for the achievement of NEC states.

Seventh, the NEC conception mirrored some general principles of the British defence acquisition system and thereby already foreclosed the implementation approach of NEC. This differs from the U.S.' conceptual framework on NCW that did not elaborate on the implementation of NCW/NCO. The 2003 DSTL document, for example, introduced "NEC Core Themes" (Defence Science and Technology Laboratory 2003), which were a list of future military capabilities, such as full information availability, shared awareness, agile mission groups, synchronised effects, effects-based planning, resilient information infrastructure, and fully networked support and flexible acquisition. At first sight, it seemed the core themes were the re-iteration of the NCW value chain with an odd attachment of support and acquisitions issues, but this was not the case. With the identification of these core themes, this document rather laid out a rough implementation scheme and carved out what – in the various sectors of defence – needed to be achieved in order to bring NEC to life. The 2005 Handbook also focused on NEC implementation, as it merged NEC with the MoD's capability management framework (defence lines of development, DLOD). The 2005 Handbook concluded that "NEC will be implemented through the coherent and progressive development of Defence equipment, software, processes, structures, and individual and collective training underpinned by the development of a secure, robust and extensive network of networks" (British Ministry of Defence 2005).

Finally, the 2005 Handbook described the governance of NEC (again, another dimension of bringing NEC to life). Although it acknowledged the defence-wide application of NEC, it focused on the implementation in the operational realm. NEC development was directed by the Joint Capability Board (JCB) led by the Deputy Chief of Defence Staff (Equipment Capability) (DCDS (EC)) and the Command and Battlespace Management Board, led by the Vice Chief of Defence Staff on behalf of the Chief of Defence Staff. Specific programme objectives were derived from the application of the High Level goals and the capability management framework DLODs. For each identified implementation cluster, a one-star general was responsible for achieving the envisaged capability. To name but a few, the clusters consisted of change objectives, such as network reach, network resilience, individual training, team training and common operational data sets. In total, the NEC handbook defined 16 such change objectives. As a result, responsibilities for NEC implementations were fragmented but at the same time hierarchically structured and clearly distributed.

The most important difference between NCW and NEC was the scope of the British concept. NEC was less ambitious than the original U.S. concept and hence, implementation was more pragmatic from the beginning. The conceptual differences in the British and the American view partly stem from different perspectives on the use of technology in war. Apparently, scepticism existed in the UK towards the use of technology in war, especially when technology was seen as a substitute for human (inter-)action.[41]

4.1.2.4 British Scepticism Towards the Promises of the Use of Technology in War

Abundant studies exist that summarise the characteristics of the American way of war (Boot 2003; Gray 2006a, c; Weigley 1977). Usually, technophilia and the focus on firepower are amid the set of attributes used to describe the U.S. war-fighting culture. During the 2000s, U.S. officers were highly supportive of information-age warfare (Mahnken 2008: 225).

The British way of war departs considerably from these instances of technocentrism. In a nutshell, there are three basic doctrinal differences between British and American military thinking that influenced the way the original NCW concept was perceived and later adopted.[42] These differences include the different perspectives on human action in war; second, the scepticism towards the assumption that technological advances automatically improve military function; and, third, a different view on the ultimate aim of military action.

From the beginning, NCW was perceived critically in the UK, as the concept did not adequately address the issue of human action and responsibility in combat (Potts 2003). In 2005, General Mike Jackson put his finger on this[43]: "The language here is quite interesting. In the United States they call it 'network-centric warfare'. 'Centric' is an interesting word there, is it not? That implies that the network is at the centre. Our view is somewhat different, because we do not quite see it that way. (...) At the end of the day, being an infantry officer (…) you have infantry soldiers at the point of decision. They will be at the centre, in that sense of decision-making. I believe our phrase, 'network enabling', is absolutely right. Better, more sophisticated, and faster communications enable us to do things we have done before (…) but much more quickly, smoothly, securely, more easily."

This focus on the individual resembles the British Defence Doctrine (British Chiefs of Staff 2001: 5-1): "Its [the British Way of War] application is denoted by constant adaptation to circumstances and a preference for empirically-based rather than theoretical

41 One military official went so far as to claim: "Brits absolutely despise the RMA" (Interview, London, 11 November 2009).

42 For a general comparison between the American and the British way of war, see Egnell (2006).

43 Quoted in the 4th Defence Select Committee Report, 3 March 2005, http://www.publications.parliament.uk/pa/cm200405/cmselect/cmdfence/45/4502.htm (accessed 15 May 2010).

or necessarily technological solutions. An emphasis on professional competence engenders an uncompromising approach to training, in order to acquire and maintain the skills necessary to prevail in the most challenging situations. However, its most distinctive feature relates to the recognition that warfare is a human activity, both in its exercise and effects, and that the most decisive element is the contribution made by men and women suitably trained, equipped and led."

A comprehensive account of this distinct way of recent British military thinking is provided in the writings of General Rupert Smith who – as the director of studies in the Staff College (Army) in the late 1980s – was the intellectual force of the British Army transformation and reportedly has "influenced an entire generation of officers".[44] General Rupert Smith commanded the British Armoured Division in the first Gulf war in 1991, the UN Protection Force (UNPROFOR) in Bosnia in 1995 and was – during the Kosovo War in 1999 – Deputy Supreme Allied Commander Europe (which is the second most senior military NATO post in Europe). The British understanding of the nature of war and the importance of command is best captured Smith's book *The Utility of Force* (Smith 2005). The central idea of *The Utility of Force* is that, due to institutional thinking and resistance to change, western militaries are still designed to fight what he called an "industrial war"; aiming at the destruction of the enemy. Yet, 20th century industrial war, Smith argued, did not exist in that form anymore. Instead, western militaries had to fight wars amongst the people. And here, the objective must not be to destroy the enemy, but to change the mind of the population.

A second difference in the underlying philosophies of NCW and NEC is the determinism with which the U.S. American NCW concept claimed that a robust network structure almost *automatically* would lead to superiority in the battlespace (see NCW value chain), which cannot be found in British military thinking. Instead, "battle is an event of circumstance, no matter how much planning, exercising and drill precede it. The chances of victory are undoubtedly increased with proper preparation, but ultimately opponents fight the battle of the day: on another day, in the same location, with exactly the same forces, they would fight another battle in different circumstances" (Smith 2005: 64).

As a result, the British NEC concept hardly ever mentioned combat superiority at all. In the British understanding, superiority could not be guaranteed by any concept or technology.

The British were, furthermore, sceptical when it came to assuming that adapting technological advances would automatically increase military performance. Here, Smith referred to the historic case of the French Army in 1870 that was proud to field the Mitrailleuse (an early kind of machine gun) but was not able to deploy it in the right way as to exploit all of its advantages (Smith 2005: 75). "It is no good acquiring technology if it is not used to advantage, which may require adapting your organisation and tactics accordingly. Alternatively, if there are good reasons not to adapt, then you should question whether it is necessary to burden yourself with the new technology. Most armies, including the British, have failed to learn this lesson – particularly in the field of modern communications."

In yet another quote, Smith told the reader: "Operational concepts and organisations tend to be adjusted to take advantage of the technology rather than to fight in a different way." (Smith 2005: 298) Specifically on the subject of NEC Smith warned of substitut-

44 Interview, London, 11 November 2009.

ing strategy for technology (Smith 2005: 400): "Great advantage is presently expected from the 'digitization of the battlefield' and 'network enabled warfare'. However, we must be careful to understand where we seek to gain this advantage and over what. There is a danger of knowing more and more about oneself and proportionally less and less about the enemy. Information technology should be harnessed to support the information operation being conducted to understand and find the opponent and separate him from the people, and to network the effects of our actions so as to complement one another." This view was resembled in the British scepticism about the substitution of human intelligence by sensors in the theatre (Le Fevre/Thornton 2003).

The third difference between the U.S. and UK approach to NCW/NEC ultimately concerned the *aim* of applying the new concept. In all of the original NCW publications, improving combat performance was always at the end of NCW. The British NEC approach, in contrast, has a more comprehensive view on security. Conflict resolution did not necessarily mean that the adversary had to be physically destroyed. Furthermore, conflict resolution did not necessarily end when the battle had been won. This thinking was based on a more balanced approach to conflict resolution that had been influenced by the past experiences of British military operations, most importantly, in Northern Ireland. Recent history is full of examples and anecdotes that might visualise this rift between American and British views on military ends. In his autobiography, British General Mike Jackson described, for example, what he called a "Cold War mentality" of then-NATO Supreme Allied Commander Europe U.S. Gen. Wes Clark during the Kosovo Crisis, who was believed to be spoiling for an all-out war (Jackson 2007: 293). The U.S. American taste for an air campaign led British General Rupert Smith (then DSACEUR) to comment on this as "an unnatural act (...) committed in a dark and unpleasant place (...) that produces nothing (...) but has a certain rhythm to it" (quoted in Jackson 2007: 305).

In sum, fundamental differences existed between British and American military thinking concerning the application of technology in war and what could be expected of it. Apart from evidence in military writings and doctrinal documents, Theo Farrell and Tim Bird were able to establish this scepticism in a survey of British military officer students (Farrell/Bird 2010). Interviews conducted for this book supported the view that this scepticism about the role of technology in war accounted for the substantial differences between the original NCW concept and the more modest NEC version and somewhat unenthusiastic reception of NCW. Shortly before the adoption of NEC in the UK, the UK Joint Vision had been issued, which nicely captureed the British understanding of technology in war and almost read as a manual on how to approach the adoption of the novel NCW concept (Joint Doctrine and Concepts Centre 2001: 502): "The British way of warfighting will not fundamentally alter, although elements of it will be developed as technology provides the opportunities to realise certain elements more fully."

4.1.2.5 Motivations for NEC Adoption and Adaptation

The process of conceptualising NEC was steered from the inside of the MoD. It was not – as in the case of Germany – partially a matter of outside entrepreneurship.[45] If it can be accepted that, in the British case, NEC is in fact the continuation of the earlier digitisation efforts, then it is presumably worth looking at the underlying motives that led to digitisation. A first trigger was the successful *demonstration* of digital C2-equipment by U.S. troops in the First Gulf War and the subsequent U.S. Army digitisation programme under U.S. Army General Shinseki.[46] In order to stay interoperable with U.S. Armed Forces, therefore, the UK started – with some delay – to go down the road of digitisation itself. The second reason was a result of the changing nature of conflicts after the end of the Cold War and the adapted design of the British Armed Forces: Smaller, rapidly deployable units made the case for a new Army digital C2 system and the establishment of a Joint Rapid Reaction Force and a Permanent Joint Headquarter required the development of a joint C2 system (British Ministry of Defence 1998).

What have been the motives for introducing NEC into the British Armed Forces? *Interoperability with the U.S. and Jointness* had been important motives already during earlier digitisation efforts in the UK and were two important drivers in the case of NEC adoption. In the NEC Handbook 2009, for example, jointness was even defined as the ultimate aim of all NEC efforts in the Network, Information and People dimensions. Yet, it is not so much jointness, but providing interoperability with the U.S. that was found to be the most important factor for the NEC programme (Chuter 2004; Farrell/Bird 2010; Sloan 2008: 57).

Interoperability with the U.S., however, was not seen as an end in itself but was justified with operational reality and the needs in the theatres of Iraq and Afghanistan. This evidence is backed-up by official statements as well. The MoD Annual Report 2003/2004, for example, points frankly at the importance of ensuring interoperability with the U.S. by saying that "it is inconceivable that the most demanding of operations could be undertaken without the involvement of the United States (either leading a coalition or as part of NATO)" (Annual Report 2003; British Ministry of Defence: 32). On that same note, the IRAQ – First Reflections Paper, states: "Equally important were the benefits of training and operating together [with the U.S.] over many years, especially in the Gulf, Afghanistan and the No-Fly Zones over Iraq. Given U.S. technological and military dominance, we should continue to track, align with and integrate U.S. developments in areas where our force balance and resources allow, particularly in terms of the organisation of enhanced HQs, communications and information systems, and Combat Identification (ID). We should also ensure that our command structures can engage and influence key U.S. decision-makers with appropriate weight and at the right levels" (British Ministry of Defence 2003b: 4). Further below, in the section "Implications for Future Joint Operations" the document goes on to say that: "The overwhelming success of rapid, decisive operations in Iraq reflects the deployment of fast moving light forces,

45 Military officers interviewed for this book in 2009 named the Vice Chief of Defence Staff, the Deputy Chief of Staff (Capabilities), the Command & Battlespace Management (CBM) unit in the MoD and various persons that held positions (DEC (CCII), VCDS) all contributing to the MoD Equipment Capability Customer. From the three individual services, only the RAF is mentioned once, with Army, Navy and the PJHQ not mentioned at all.

46 Interview, Warminster, 16 June 2009.

highly mobile armoured capabilities and Close Air Support, which made use of near real-time situational awareness by day and by night. The U.S. ability to combine land and air operations and support them from the sea and from friendly bases at very high tempo enabled the mix and impact of joint assets to be adjusted to operational need or events across the whole theatre of operations. This is likely to shape U.S. doctrinal development and impact on potential partners. The implications of maintaining congruence with an accelerating U.S. technological and doctrinal dominance need to be assessed and taken into account in future policy and planning assumptions" (British Ministry of Defence 2003b: 4).

Achieving and sustaining jointness and interoperability with the U.S. have been found to be the two most important reasons for the introduction of NEC. Had other potential motives discussed above in theoretical framework been present? The following paragraphs will briefly discuss the other potential drivers for NCW adoption, notably national ambition, the role of the U.S., the role of NATO, other states acquiring NCW, domestic politics, mission orientation (operational need) and industrial lobbying.

Military officials interviewed for this book have hardly ever mentioned *national ambition* as a motive for NEC. Also, no evidence could be found for an active *push from the U.S.* (the inventor) in the case of British concept-adoption. Instead, evidence shows that a) it was the UK that actively requested information on NCW and b) that the U.S. entered an exchange of ideas with the UK on the subject early on. With a view on operational interoperability (bearing in mind the coalition operation in Afghanistan and later on in Iraq) it was the UK striving for interoperability rather than fulfilling U.S. demands. Yet, Theo Farrell and Tim Bird argue that a more subtle emulation mechanism might have been at work stemming from Britain's close military proximity to the U.S. (Farrell/Bird 2010: 36).

Also, the *role of NATO* in spreading or utilising the NCW concept for the UK was not a determining factor (Farrell/Bird 2010: 55). In fact, it seemed to be the other way around: It was the British that actively shaped the NATO approach of NATO Network Enabled Capabilities (NNEC) that appeared between 2003 and 2005.

NCW adoption in other states did influence British NEC introduction. The Swedish success of Network-Based Defence (2001/2002) apparently influenced the British implementation of their own concept. Furthermore, as part of ABCA, the Australian attempts to develop their own NCW force justified and pushed British efforts in this area as well. But despite functioning as positive examples to justify British efforts, no other mechanisms (neo-realist or isomorphic) could be singled out during the interviews.

Did *domestic politics* play an important role? Unlike many other NATO states, the UK did not proclaim a *military transformation*. According to one military official interviewed for this book the British despised the notion of a *defence transformation.*[47]

Hence, NEC did not function as a motor of or a justification for reform. Another domestic reason apparently bore some importance: saving money. The cost-saving aspect of NEC was openly addressed in the 2003 Defence White Paper. Among other advantages, NEC was seen as an enabler to allow "the same military effect to be achieved

47 As the idea of a *military transformation* had originated in the U.S. and the British were rather reluctant adopt the very same notion all the more because "we think we know best" (Interview, London, 30 October 2009). Also see the article by Michael Codner on this topic (Codner 2005). ther observers still applied this notion when analysing the recent military changes in the UK defence system (Farrell 2008).

with less" (British Ministry of Defence 2003a). This view – criticised for sharing ideological ground with the Rumsfeld doctrine of substituting cost-intensive manpower with technology – has somewhat been changed over the following years and, today, the MoD position does not reflect saving only but rather the belief that the NEC initiative would contribute to more coherent acquisition, interoperability between purchased platforms and congruency between projects.[48] Finally, NEC has been placed in the context of contributing to the Comprehensive Approach (CA) by enabling information exchange not only between military chains of command, but also across different departments.[49]

The *industry* was less involved as an external entrepreneur trying to place new (NEC) products, but was rather a partner in bringing the British NEC vision to life. The big and established defence industrial companies were rather ignorant towards NEC and small- and medium-size enterprises filled niches in the realm of information management technologies.[50]

To conclude, NEC was very much about pragmatically improving the British way of war and was shaped by the operational reality in Afghanistan, and later Iraq. Other potential motives, for example industrial lobbying, tended not to play a major role in NEC utilisation.

4.1.2.6 Conclusions

The tracing of the NEC-utilisation process and the comparison of the British NEC and the American NCW concepts offered some insights into the differences between both concepts, but also shed light on the factors that account for these differences. The pace of NEC conceptualisation was rather quick: The NEC Outline Concept was issued in 2003, only months after the NEC appeared in the Defence Review: New Chapter 2002. This quick drafting of the concept can be explained by the fact that the then JBD programme was stuck and that there was a *window of opportunity* for the new NEC concept to present a fresh start. Equally important, institutional structures in the MoD Capability Customer had already been in place. Knowledge on earlier digitisation attempts and on NCW was centralised within the MoD. The nature of this factor resembles both institutional legitimacy and efficiency. A further factor for the quick conceptualisation of NEC was the participation of the British Forces in the Afghanistan and Iraq campaigns, where they were granted access to exploit U.S. networks and, therefore, needed to keep up with American technological development – a factor resembling military effectiveness. A further effectiveness-related factor was the need to provide inter-service jointness – especially in the Permanent Joint Headquarters (PJHQ) – in the face of the ongoing operations. Finally, NEC was seen as a means of reducing procurement costs; in the face of the decreased UK defence budget in the late 1990s, this can be interpreted as an efficiency-related factor.

The comparison of NCW and NEC revealed that there were substantial differences between the two concepts in the areas of ambition, desired end-state (superiority vs. enhancement), inherent logic (domains vs. dimensions) and applicability (operational-

48 Interview, London, 18 November, 2009.

49 On the origins and the features of the British CA, see "The Comprehensive Approach", Joint Discussion Note 4/2005 and Joint Warfare Publication 3–50 "The Military Contribution To Peace Support Operations", 2nd Edition, June 2004.

50 Interview, London, 10 November 2009.

tactical vs. defence-wide). In the case of the UK, the adapted version of NCW differed widely from the U.S. original concept, therefore concept faithfulness is considered to be low. Two factors could account for this low concept faithfulness. One factor were the different views on the use of technology in war in British and American military thinking, which was mainly based on the different military cultures of the two nations. Yet, it would be too simplistic to limit these differences in military thinking purely to military culture. As the NEC concept has been adapted so as to best resemble current British military doctrine, this includes a notion of effectiveness as well. The other factor were the limited financial resources that narrowed the scope of NEC.

Table 7a: The Utilization Pace of NEC in the UK

Pace of Utilisation	Quick
Explanatory Factor	Nature of the Factor
JBD Programme Stuck between Different Stakeholder Interests	Institutional Legitimacy
Existent Institutional Awareness for NCW and Structures in the MoD	Efficiency and Institutional Legitimacy
Need to Provide and Maintain Interoperability with U.S. Forces during Operations	Effectiveness
Need to Provide and Maintain Jointness	Effectiveness
Means of Saving Money	Efficiency

Table 7b: UK NEC Concept Faithfulness

Concept Faithfulness	Low
Explanatory Factor	Nature of the Factor
Different Views on the Use of Technology in War	Military Culture and Effectiveness
Limited Resources	Efficiency

4.1.3 Implementing NEC in the UK

The 2005 Handbook offered a rough implementation scheme that structured NEC implementation along three so-called 'maturity states'. A sub-divisional plan – the NEC Capability Plan – further detailed this three-stage proposal in 2006 by adding concrete time frames to these maturity states. The initial state was to be achieved by 2009 (later pushed to 2012), the transitional state by 2015 and the mature state by 2025 (Ebbutt 2006). In the initial maturity state improvements in operational capability were expected (James/Meyer 2009). In the transitional period (2012–2018), medium-term improvements in operational capability through doctrinal and procedural change as well as through new equipment were the goal. The mature state was deliberately not hammered out by the MoD, as technological advances or operational need might turn any such plan obsolete (Argent-Hall 2009). After NEC took off it soon became one of the MoD's highest priorities for future investment (see 2005 report, Defence Science and Technology Laboratory 2002–).

4.1.3.1 NEC Projects

The technological foundation for achieving NEC in the UK was an all-embracing network similar to the Global Information Grid in the U.S. (Ebbutt 2006). To provide this network, a number of initiatives were set-up, the most important of which were the Defence Information Infrastructure Future Deployed (DII FD), the Skynet 5 (satellite communications),[51] the Cormorant (theatre-level communications) from EADS, the first instalment of Falcon (operational level communications) from BAE Systems' Integrated System Technologies (INSYTE) and Bowman (tactical communications) from General Dynamics UK (Keymer 2009). Beside the purchase of these systems, a phased-in approach was chosen to eventually interconnect Bowman, Falcon, Cormorant and Skynet 5. These systems present the backbone of the British military command, control and communications (C3) network.

While these C3 projects aimed to deliver the network hardware, in parallel a number of software applications were delivered, too: the initial Joint Operational Picture (JOP), the Joint Command and Control Support Programme (JC2SP)[52] and the Joint Effects Tactical Targeting System (JETTS). Another area of delivering NEC during the initial phase included sensors, such as the Soothsayer joint electronic surveillance system, the Reconnaissance Airborne POD Tornado (RAPTOR) and the short-range unmanned aerial vehicle Watchkeeper.

The services pursued their own NEC projects as well. With FRES (Future Rapid Effects System), the British Army, for example, tried to build a medium-weight vehicle fleet similar to the U.S. Army Future Combat System (also accompanied with structural changes towards medium-weight brigades (Dorman 2007)) – packed with command and control, communications, computers, intelligence, surveillance, reconnaissance (C4ISR) systems. Yet, because costs were running too high, the MoD put large parts of the project on hold in 2008 (Keymer 2009: 781). A further large Army project was FIST – the Future Infantry Soldier Technology Programme. FIST provided kits for individual soldiers to integrate them into the networked force. Among other things, the kits will comprise of C4I capabilities, targeting and navigation tools. The Royal Air Force (RAF) pursued their own Network-Enabled Air Capabilities (NEAC) programme. One important RAF project was the Reaper unmanned aerial vehicle (UAV), which had been purchased as an urgent operational requirement and had been deployed in Afghanistan in October 2007. Reaper could carry weapons but could also be equipped with ISR for reconnaissance and target acquisition missions.

NEC implementation was closely linked to NEC conceptualisation. This was remarkable given the plethora of equipment projects. Implementation was monitored by a single institutional entity – the NEC Programme Office (see below) that tried to retain overall consistency of the many big and small NEC projects.[53]

In the network dimension, most programmes were equipment projects in ISR and C3 (Defence Information Infrastructure, Skynet 5, Joint Command and Control Support Programme, Bowman). Remarkably, engagement capabilities such as weapons were not considered to fall under the network dimension. Most British ISR assets, including drones, reconnaissance-pods and radar systems, were already in service. Many of them,

51 Skynet was an interesting case of 'lessons learned'. During the Falkland conflict in 1982, the UK was forced to use U.S. satellite services. This lack of capability revived UK plans for their own satellite communications system, leading to a continuation of the Skynet programme. As a result, Skynet 4 was launched in 1988.

52 JC2SP integrates the Royal Air Force's Command and Control Information System, the Royal Navy's Command Support System and the Army's Joint Operations Command System.

53 The initial state timeline was presented on a special internal web-based database by the NEC Programme Office (accessed 27 May 2009).

such as the Rover receiver that allowed ground forces to view full-motion video from manned or unmanned air vehicles or the Mini-UAV or Tactical-UAV, were ordered as an Urgent Operational Requirement for the deployment in Iraq or Afghanistan (Keymer 2009: 784).

In the information dimension, the emphasis was on Information Management. In 2005, the MoD released a defence-wide Information Management (IM) Policy for Defence (British Ministry of Defence 2005), providing policy objectives and guidance. A year later the Chiefs of Staff released the Joint Doctrine Note on Information Management recurring especially on the operational dimensions of IM (British Chiefs of Staff 2006). In terms of projects, a joint reference information database (JRMIC) was planned to make sure that future applications use the same symbols. Yet, due to funding problems, the project was put on hold.

The EA/SOA capability is also worth mentioning. Developed in the U.S. and used by the DoD since the mid-1990s, Enterprise Architecture/Service Oriented Architecture wass a methodology for systems development. It was used in operation Herrick in Afghanistan for casualty tracking and assessing hospital efficiency.

Initially, and in contradiction to the NEC concept, the people dimension did not receive the same attention as the network and information domain. That was a consequence of the earlier JBD digitisation project that had brought about MoD structures reflecting a strong technological orientation. Despite the conceptual emphasis of the people dimension the importance of it was only slowly recognized in the implementation phases. Between 2003 and 2007, the only noticeable achievement was the issuing of a NEC Competence Framework (Annual Report 2005, British Ministry of Defence), which codified "the skill set needed by MOD staff to exploit Network Enabled Capability, and covers both IM and Information Exploitation" (British Ministry of Defence 2005). In 2006, training needs were analysed to derive conclusions for NEC education across defence (Annual Report 2006, British Ministry of Defence). Yet, the MoD did not sufficiently resource the NEC people area until 2007. Especially the training dimension was underfunded and technology was delivered directly into the field and not into training locations. Moreover, in the early stages of communication equipment development, close collaboration between the developers and the users was missing. Operational commanders hence complained about the insufficient C2 training prior to troop deployment. As a result, the Defence Technology Centre (DCC) was established within the Defence Academy at Shrivenham and opened on 1 April 2007. It provided the facilities and resources for mission-oriented education and training with real military C3 hardware.[54] These training facilities were partly financed as an Urgent Operational Requirement in connection with the Overtask programme; OverTask provided a C2 system that was interoperable with the NATO networks and provided theatre situational awareness, communications and mission planning down to the brigade level.

Some of the most prominent NEC programmes such as Bowman were set up well before the NEC initiative. Yet David Ferbache answers the questions that naturally begs itself, "Is NEC just digitization by another name?" by pointing out that, even though NEC was borne out of earlier digitisation attempts, its key challenge would be "to harness these systems and to understand the opportunities to integrate and optimise military capability, while continuing to invest in the future enablers" (Ferbrache 2003: 105).

54 See the website of the DCC: http://www.da.mod.uk/colleges/cmt/technology-division-1 (accessed 16 May 2010).

This NEC goal had not been fully met. While the delivery of the network components was generally satisfying, organisational challenges remained and it could be argued that the 'human dimension' was still underdeveloped (Skinner 2006). Jim Storr, former Army officer and now an independent defence analyst, also pointedly claimed that largely "the digitisation vision has failed". He found that, for a number of what he calls 'human' reasons, including organisational, social, cultural and political reasons, the vision of a digital battlefield never came through and that the new systems, such as Bowman, are overwhelmingly used in a way that previous equipment has been used before (in the Bowman example, secure voice only) thereby not adding any value to NEC (Storr 2009).

4.1.3.2 NEC Governance

The NEC initiative did not only result in the roll-out of sophisticated C4ISTAR equipment, it also impacted MoD governance structures and the acquisition process. In the initial years, the NEC project was embedded in the same structures that were responsible for the earlier digitisation attempts. As a result, NEC was largely located in the acquisition part of the MoD central staff. Responsibility lay especially with the Director of Equipment Capability DEC (ISTAR), the DEC (CCII), and the Capability Manager Information Superiority CM (IS). After the first few years, two problems became apparent. First, coherence between different C2 projects was missing. Second, the operational experiences revealed a lack of coherence in terms of adequate training and information management – something NEC had originally aimed for. Eventually, a revised NEC governance structure was set-up in 2007 and greater emphasis was put on the inclusion of the people and information dimensions. Two important structural changes have been made: a single responsible owner for NEC had been established and an NEC Programme Office had been set-up.

The Deputy Chief of Defence Staff (Equipment Capability) was assigned the role as the senior responsible owner for NEC in May 2007. His task was to ensure NEC implementation and ultimately he was accountable for the success of the programme. The NEC Programme Office was formed based on the existing Command and Battle Management (CBM) programme staff. The NEC Programme Office supported the Senior Responsible Owner by setting up and overseeing an NEC Campaign Plan for the MoD. It monitored and supported the work of the programme managers and also acted as a contact point for the wider defence community.

Other aspects of NEC governance included collaboration on NEC issues with other militaries and the industry. The MoD established a standing exchange forum with the U.S. on the matters of NEC equipment. In early 2002, a UK/U.S. Interoperability Commission was set-up to solve interoperability problems against the background of current operational needs. The U.S. was represented by the Under Secretary of Defense for Acquisition, Technology, and Logistics and the Assistant Secretary of Defense for Network and Information Integration, while the UK was represented by the Deputy Chief of Defence Staff (Capability) and his staff. The Interoperability Commission was concerned with a number of projects that lay at the heart of the UK NEC implementation: MNIS Command Systems Interoperability (CSI), the interoperability between the GIG (U.S.) and the GII (UK), between WIN-T (U.S.) and Falcon (UK), between JTRS (U.S.)

and Bowman (UK), between JSTARS (U.S.) and ASTOR (UK) and – amongst others – focused on more functional issues like joint fires or the single integrated air picture.[55]

There was also NEC-related collaboration between the MoD and the defence industry. Apart from public-private partnerships in the areas of maintaining systems and services (i.e. Skynet), collaboration in the realm of research and development increased as well. In 2002 an NEC Delivery Team was established to contribute to the utilisation of NEC and to assess the impact on the DLODs. The Delivery team was led by the DSTL with the support from Qinetiq and other industry representatives. A year later, in 2003, NITEworks was set up – a joint project between the MoD and the industrial partners (BAE Systems, EADS, GD, LogicaCMG, MBDA, QinetiQ, Raytheon, Thales and Finmeccanica) that operated mainly in the realm of Concept Development and Experimentation and was a major initiative of the MoD to boost the development of NEC capabilities by harnessing the research capabilities of the industry (Jones 2008; Skinner 2008).

4.1.3.3 The Application of NEC

Apart from NEC projects and governance, it is worth looking at the actual application of NEC in operations in order to assess NEC implementation. One has to bear in mind, however, that the NEC programme is only in its first phase and that delivery of NEC remained limited still.

In 2008, Jane's International Defence Digest published an article entitled *UK forces in Afghanistan Begin to See NEC Dawn* (Ebbutt 2008). It was argued that the realities of coalition operations in Afghanistan were ultimately beginning to shape the UK's NEC programme. This was due to the positive pay-offs after the UK had decided to abandon the longsome development of JOCS (Joint Operational Command System) in favour of the OverTask network in 2007. Two years earlier, in 2005, two other important systems that contributed to NEC capacity entered into service – the Skynet satellite and the Bowman communication system (Skinner 2006).

As NEC was still in its initial phase, however, no overall network existed. Instead, a number of different C3 networks (with different classification levels) were operating in parallel. The famous swivel chair interface between C2 systems still existed and the British operations room in the ISAF HQ in Helmand province reportedly had 16 different situational awareness systems and 25 different information systems in 2009 (Temperley 2009). In the case of Iraq, early observers saw "UK/US interoperability on track" (Kemp 2003), while later commentators concluded rather that cooperation nearly failed in the first phase of the invasion due to interoperability and communication problems (Skinner 2006).

There were a number of challenges with regards to NEC that became apparent during the operations in Afghanistan and Iraq. Those challenges were to be found less in the network domain, but in soft issues, notably training, information management and cultural resistance against NEC.

According to a study by the Pentagon's Office of Force Transformation examining the use of blue force tracking (BFT) by both U.S. and British forces in Iraq, BFT and satellite communication capability did provide commanders with a common operational picture (Garstka et al. 2006). Yet British troops claimed that, due to insufficient training, they did not trust the technology and therefore failed to use it to its full potential.

55 Personal email communication with NEC PO on 20 August 2009.

The authors of the study furthermore concluded that BFT information was rather suitable for the operational and strategic command levels but not so much for the tactical level.

Apart from insufficient training, there was a more general lack of qualified personnel. Due to a Report by the House of Commons Defence Select Committee the Army was short of UAV operators (there was a 50% shortfall in 2006 and in 2008). The situation of imagery analysts revealed a similar picture. Moreover, the introduction of Bowman caused significant problems for the Army, as it lacked sophisticated personnel to run Bowman; "There is a difference between repairing radios and managing a networked environment".[56] The army tackled this issue by putting greater emphasis on engineers in a review of its structure (Future Army Structure). In 2009 the MoD announced additional ISTAR capabilities including more flying hours for UAVs to support troops in Afghanistan (Keymer 2009).

During operations, the relation between technology and human actions also appeared problematic. At the sensors-level, the UK successfully deployed UAV[57] and Tornado Recce-Pods and delivery of equipment in the realm of ISTAR during the last few years was quick and targeted. The Defence Select Committee acknowledged "from the experience of current operations, the MoD is broadly content with the assets it has, such as UAVs, which collect ISTAR information."[58] According to a study that assessed the overall application of NEC by the UK forces in Iraq after the invasion, however, commanders were reluctant to rely on all available ISTAR assets because they were afraid of indecisiveness or command paralysis. There was doubt among officers that NEC was suitable for low intensity operations (Griffin/Whetham 2007: 235). Low intensity operations (LIO), according to the officers' views, should rely more on human intelligence – that is, the British soldier on the ground – than on ISR technology like UAVs.

A further problem in operations was information management. In 2008 the Defence Select Committee pointed out that "improvements are required in relation to the Direct, Process and Disseminate elements of the ISTAR chain".[59] Deployed headquarters still lacked the capability to analyse and link existent information so to provide that kind of situational understanding that NEC hopes to deliver. Griffin and Whetham have also made this point in their case study on NEC in Iraq (Griffin/Whetham 2007). They concluded that "there are concerns about the UK's ability to fully utilise its existing ISTAR assets. The issues require further analysis, but some argue that the UK's capacity to analyse and integrate the information acquired through ISTAR is currently outstripped by the capabilities themselves. Better and faster analysis of such information may require some organisational restructuring with enhanced NEC at [brigade] and [division] levels, notably the expansion of their intelligence infrastructures" (Griffin/Whetham 2007: 236).

Another observer points out that there is "little point in acquiring yet another intelligence gathering asset when the UK's existing information management capabilities within ISTAR system were unable to effectively make use of the data gathered by

---

56 Interview, London, 2 December 2009.

57 www.mod.uk/defenceinternet/defencenews/equipmentandlogistics/unmannedaerialvehicle sonthelookok outoveriraq.htm (accessed 6 February 2011).

58 See "Defence- Twelfth Special Report", House of Commons, 21 October 2008, http://www. publications.parliament.uk/pa/cm200708/cmselect/cmdfence/1087/108702.htm (accessed 6 Fe-bruary 2011).

59 See footnote 58.

them" (Downs 2009). Turning information into knowledge clearly falls into the Information Dimension. Although – with the exception of sufficient bandwidth[60] – the MoD successfully delivered network components into the field, it apparently failed to provide sufficient means for information and knowledge management.

Finally, there was some neglect to acknowledge the value of NEC. In their study about the application of NEC in Iraq, Griffin and Whetham quote a General who gives an impression of this: "I have a little test I give operational commanders. When I talk to them in theatre I ask them how much bandwidth they have. They all know how much fuel they have, or how many tanks, but many don't have a clue how much bandwidth they have or how to use it efficiently" (Gen. Ridgeway quoted Griffin/Whetham 2007: 236).

This neglect could become problematic, for example, if commanders or staff officers do not feed information into the command systems to the effect that others cannot access them.[61] In the case of Bowman, users continued to work with Bowman as if they had their old systems at hand, therefore not only failing to fully exploit the system but even creating undesired effects.[62] There were reportedly incidents in which deployed commanders desperately tried to download power point presentations from an MoD site thereby blocking precious satellite capacities for hours.[63]

In sum, the UK managed to field an initial NEC capability to the Afghanistan and Iraq theatres. Sophisticated C4ISTAR equipment had been purchased as UOR and fielded quickly. On the other hand, the full exploitation of the NEC opportunities was problematic as sufficient training capacities, personnel, information management and an overall cultural awareness was missing in the beginning of the NEC project. The redirections in the programme in 2007, however, addressed these deficiencies.

4.1.3.4 Conclusion

Implementation pace and faithfulness of implementation were the final variables by which to assess the adoption of NEC and to make assumptions concerning efficiency/ effectiveness and institutional legitimacy as drivers for NEC: The NEC implementation process was quite transparent. On the one hand, the MoD pursued a rather open communication strategy for NEC. In the NEC Initial State Timeline, for example, potential drawbacks and delivery delays were openly addressed. The NEC handbooks, furthermore, were not only open to the public but meant to be the basis for raising pan-Defence NEC awareness.

A second observation is that through the DLOD structure and after the reform of the NEC governance structures in 2007 the technical and non-technical NEC features were integrated to form a comprehensive NEC implementation approach. Third, implementation clearly had an operational focus, especially ensuring interoperability with the U.S. military. An internal communication from the NEC Programme Office stated an order of the interoperability priorities – joint, U.S., NATO, EU and then other partner nations. For example, the priority in developing the multinational information sharing MNIS system (a set of applications allowing secure communications such as chat rooms be-

60 Interviews, London, 18 November 2009 and 2 December 2009.
61 Interview, London, 11 November 2009.
62 Interview, London, 2 December 2009.
63 Interview, London, 2 December 2009.

tween coalition partners) was to achieve interoperabilty with the U.S. GRIFFIN network. The long-term naval CEC project also focused on the full C2 interoperability of UK and U.S. navies. Another example was the ASTOR airborne radar system, which was interoperable with the U.S. JSTAR and the U.S. Common Ground Station. The Reaper drone is another good example. This aerial vehicle was remotely operated; while landing and take-off were operated from ground units in the theatre, the actual mission command was exerted via satellite communication from a crew stationed in Creech, Nevada.

The fourth observation is the significant pragmatism with which NEC delivery was pursued. Many projects had been purchased as UOR to quickly field required equipment (Ackerman 2009a). Many of these UOR were radios and communication devices along with the required software. Another example was the Tactical UAV solution Hermes 450 that was procured to bridge the gap between the out-of-service Phoenix UAV and the Watchkeeper UAV, which was not in service yet. Another example was the network capability J1/J4 IOS (Interim Operations Support) in Afghanistan that was initiated following the delay of a DII increment in 2005. Generally, UOR were seen as important complementary ways to bring NEC to life beyond the allocated budget. One drawback, however, is the potential of undesired course changes in the NEC programme through UOR.

Fifth, and on a more negative note, the acquisition process suffered from underfinancing and delays. Under-funding partly stemmed from a generally tight defence budget and the strenuous military operations in Afghanistan and Iraq. Yet, unlike in the U.S., the major NEC programmes had been fully funded earlier in the procurement cycle and thus were not terminated. Yet this was not true for smaller projects, projects in the development phase, follow-up projects and projects not considered as urgent – some of them have been cancelled (Ackerman 2009a).

Some funding problems occurred because programme-planning assumptions still tended to rest on the idea that projects remained in a steady state of development. C2 systems were especially subject to continuous technological change and rapid developments in civilian digital communications, but also to operational and interoperability challenges. This problem occurred, for example, in the case of the Bowman radio (UK National Audit Office 2006) and the Typhoon Future Capability Programme 2, which sought to provide Link-16 capability for the Typhoon aircraft. The Soothsayer project, to give another example of a severe delay, was heavily criticised in the 2008 Major Project Report from the NAO. Apart from an immense cost-overrun (52 million GBP), the programme was then delayed by two years. As a result, the MoD decided to terminate this and other problematic projects in 2009.[64]

A final observation is the extensive application of Public Private Partnerships, Through Life Management and commercial/military-off-the-shelf products (C/MOTS) in the NEC programme. The Hermes UAV, for example, was partly operated by Thales/Ebit civilian pilots that controlled take-off and landing. Skynet, the communication satellite capability, was operated by Paradigm Secure Communications on behalf of the UK Ministry of Defence. The Soothsayer programme relied on commercial-off-the-shelf products.[65]

64 See *UK MoD pulls plug on Soothsayer*, Jane's Defence Weekly, 29 May 2009, p.7.
65 See http://www.lockheedmartin.co.uk/news/167.html (accessed 26 October 2009).

Implementation of the first, initial NEC stage, proceeded rather quickly. Especially in the network dimension (the hardware), a number of projects – Bowman, Skynet, UAV and Overtask – have successfully been delivered. In the case of Bowman, the programme was even largely re-adjusted between 2000 and 2003 in order to meet operational requirements and fit the new NEC concept (UK National Audit Office 2006: 8ff.).

Implementation did not progress steadily. By 2007, the NEC programme appeared to be unbalanced. The people and information dimensions were neglected to some extent. Experiences from deployed troops revealed insufficient training on newly fielded communication devices. Moreover, commanders did not make full use of the available equipment.

Table 8a: The Implementation Pace of NEC in the UK

Pace of Implementation	Quick (Moderate)
Explanatory Factor	Nature of the Factor
Many NEC-Relevant Equipment Projects were in the Pipeline	Efficiency and Institutional Legitimacy
Quick Roll-Out of C4ISTAR Equipment under UOR for Operations in Afghanistan and Iraq	Effectiveness

Pace of Implementation until 2007	Moderate
Explanatory Factor	Nature of the Factor
NEC Governance through Existing Equipment-Oriented MoD Bodies	Institutional Legitimacy

Pace of Implementation since 2007	Quick
Explanatory Factor	Nature of the Factor
Re-Adjustment of People and Information Dimension after Experienced Ineffectiveness of Deployed C4ISTAR in Operations	Effectiveness

NEC governance, furthermore, was initially conducted by the same institutional entities in the MoD Capability Customer that had been previously concerned with the digitisation attempts prior to NEC. This resulted in a strong focus on equipment-delivery while too little attention was paid for the other DLOD dimensions. This fact is a strong indicator for path-dependent behaviour and therefore for increasing institutional legitimacy (relying on established structures to maintain support from other actors in the defence system) before 2007. It is well established in the literature that pre-exiting patterns and routines in an organisation shape the way new problems are perceived (Allison/Zelikow 1999: 149ff.; March et al. 1993: 170f.).

Table 8b: UK NEC Implementation Faithfulness

Faithfulness of Implementation	High (Moderate)
Explanatory Factor	Nature of the Factor
Mission Orientation	Effectiveness
Quick Roll-Out of C4ISTAR Equipment under UOR for Operations in Afghanistan and Iraq	Effectiveness

Faithfulness of Implementation until 2007	Moderate
Explanatory Factor	Nature of the Factor
Preference for Deliveries in the Network Dimension (Equipment Orientation)	Institutional Legitimacy

Faithfulness of Implementation since 2007	High
Explanatory Factor	Nature of the Factor
Re-Adjustment of People and Information Dimension after Experienced Ineffectiveness of Deployed C4ISTAR in Operations	Effectiveness

In a reaction to the apparent problems of NEC, Elizabeth Quintana – head of RUSI Military Information Studies – provocatively asked: "Is NEC dead?" (Quintana 2007). Her paper was not well received within the MoD and some months later, Air Vice-Marshal Stu Butler (Capability Manager for ISTAR in the MoD) felt obliged to reply in RUSI Defence Systems that NEC in fact was "Alive and Well in the MoD" (Butler 2008), focussing his evidence especially on the new governance structures with a Senior Responsibility Owner and the devotion to Through Life Management for NEC. In fact, many of the existing problems were alleviated with the setting-up of the new NEC governance structure in 2007, the set-up of new NEC training facilities and the greater efforts in the information dimension. The table above illustrates the findings on the NEC implementation speed in the UK that differentiate between the two periods of moderate and high implementation speed.

Turning to the faithfulness of NEC implementation reveals a similar picture. The overall faithfulness of implementation is high – this shows the appointment of responsible owners for NEC delivery along the DLOD matrix, the reflection of the dimension model in the implementation programme and the NEC governance structure that also reflects the dimension model. Until 2007, the information dimension and the people dimension, however, had received less attention than officially proclaimed. As a result, implementations faithfulness can only be considered moderate until 2007. The main reason for the overall high level of faithfulness of implementation was the mission orientation of the NEC project and therefore its implementation. In order to be interoperable with the U.S., a number of C4ISTAR projects had been pushed forward, partly financed as UORs. After the adjustments of the NEC governance structures in 2007, NEC faithfulness of implementation appeared even higher.

4.2 NEC in the UK: Concluding Remarks

The aim of this NEC case study was to assess the adoption pace and the adoption patterns. In a second step these findings were linked to explanatory factors that were either efficiency/effectiveness-oriented or institutional legitimacy-oriented in nature and to figure out to which level – the international, the domestic, the organisational – the military organisation was most responsive to in the course of NEC adoption.

With a view to the more descriptive task of establishing the unique adoption patterns it could be shown that the UK NEC programme started early in 2002. In comparison to other cases of NCW adoption the UK (along with Sweden) was one of the earliest adopters of the NCW concept. Similarly, the pace with which knowledge acquirement, conceptualisation and implementation proceeded was rather quick (with some qualifications during the knowledge acquirement and implementation phases). UK adoption patterns could therefore be described as follows:

Table 9: NEC Adoption Patterns in the UK

Timing	Early
Adoption Pace	Quick (Moderate)
Concept Faithfulness	Low
Faithfulness of Implementation	High (Moderate)

The second aim of the research was to find explanations for the specific adoption patterns. While in the concluding chapter of this book the findings of the British case study will be discussed against the backdrop of German case study results some tentative conclusions can already be drawn without explicit comparison of both cases.

How can the NEC adoption patterns in the case of the UK best be explained? How do the findings (the adoption patterns) relate to the presence of a specific set of external factors? As qualitative research revealed it were mostly efficiency/effectiveness-related factors (in contrast to institutional legitimacy related factors) that accounted for the swiftness and relatively high faithfulness of implementation of the NEC adoption. The drivers for NEC adoption in the UK are summarized in the following table. Double lining highlights what has been found to be most critical triggers for British NEC adoption.

Table 10: Drivers of NEC Adoption in the UK

Level	Efficiency/Effectiveness	Institutional Legitimacy
International	Need for Interoperability with U.S. and Coalition (Incentive) Operational Requirements (Incon tive)	Ambition/Prestige (Incentive)
Domestic	NEC as a Means to Save Resources (Incentive)	
Organisational	Need for Joint Interoperability (Incentive)	Pragmatic Approach towards Innovations (Incentive) Pre-Existing Institutional Structures (Incentive and Constraint)

As the table shows these factors stemmed mostly from the international level – the coalition requirements and the operational environment although domestic demands (budget pressures) also played a role. The most important of these factors – indicated by double-lines in the table – are the need for interoperability with the U.S. Armed Forces during the operations in Afghanistan and Iraq, the continuous need for inter-service jointness and, more generally, the timely provision of C4ISTAR-equipment for deployed troops (operational requirements). This finding is captured in a quote by RN Commander Kevin Shaw: "Current operations have had a dramatic effect on the development of NEC for the UK." (Shaw/Ebbutt 2009)

The low concept faithfulness – the degree to which NEC resembled the original U.S. NCW concept – could be understood as both being a result of effectiveness and legitimacy-related factors. Undoubtedly, different doctrinal military thinking about the use of technology in war is part of a military's culture and would thus influence perceptions about appropriate behaviour. Nevertheless, it would be too simplistic to reject out of hand any notion of effectiveness when explaining low concept faithfulness: British NEC conceptualisation had been oriented on current British military doctrine; it reflected operational experience and the likely mission spectrum of the British Armed Forces (British Chiefs of Staff 2001, 2008). In the set of questions on military effectiveness that Millett's et al (1986) assembled, one question reads: "To what degree does the operational doctrine of military organisations place their strengths against their adversary's weaknesses?" In British military doctrine, a less technology-focused approach was favoured because it was perceived as being more effective in 'low-intensity' operations.

A further yet less decisive factor for the quick and forceful pursuance of the NEC project, especially in the beginning, were the expected savings due to an effects-based approach to operations.

Of course, institutional legitimacy was also present during NEC adoption although less prominent than effectivess/efficiency-related motives. Evidence from interviews suggested that there was awareness that other states, especially Sweden, quite forcefully implemented NCW as well. It could not be established during the research to what extent national ambition or the quest for prestige – being one of the early adopters – helped the NEC programme to attract support. Presumably it did lead to a positive attitude and thus worked as an incentive.

Institutional legitimacy also worked its way at the organisational level. British Defence often hailed the pragmatism (in contrast to dogma or ideology) by which the British way of war is characterised (British Chiefs of Staff 2008: 5-1, also see Avant 1994). Without speculating about other domains of the MoD portfolio, one can conclude that NEC introduction was pursued with a significant amount of pragmatism, too. A telling example is the re-adjustment of the NEC programme in 2007. Admittedly, NEC implementation in the beginning was uneven, focusing too much on equipment thereby paying too little attention to human resources and information management. Yet, in retrospect the re-launch of the programme including the change of government structures just showed the dedication to make NEC a success. Thus, the reaction to the shortcomings of the NEC programme by readjusting implementation and structures NEC in 2007 could even be regarded as a text-book case of lesson learning in a successful policy cycle (Easton 1965).

A second area in which institutional legitimacy influenced NEC adoption concerned structures. Here, influence of the organisational level was mixed being an incentive and a constraint at the same time. On the one hand the existence of awareness for the importance of information technology for the armed forces in the MoD due to the experiences in the First Gulf War and the resulting joint digitisation attempts prior to 2002 paved the way for the quick introduction of NEC. This can also explain the timing of the decision to go 'NEC' – when the JBD programme was stuck between different stakeholder interests. Presumably, the existence of established institutional structures for the delivery of a joint command and battle management system within the MoD eased the introduction of NCW in Britain. On the other hand, however, these established structures proved to be an obstacle for quick NEC implementation later, as they were focussed too much on the delivery of C4ISTAR-equipment and too little on the non-technical issues of the programme. Yet it has been discussed in the previous paragraphs that the NEC programme was adaptive and that in 2007 the programme was re-adjusted.

When asked why the UK was so much faster in bringing NEC to life than other countries, one former official in the procurement area explained: "If you keep on thinking for too long you are going to lose momentum".[66] Increasing efficiency and effectiveness of the British Armed Forces in military operations were the main factors that contributed to the early timing, the fast pace and the high faithfulness of implementation of the introduction of NEC. In sum, in the UK the NEC project was adapted so as to best suit current missions and its implementation was pragmatically pursued so as to quickly bring about the expected results. Thus, in the case of NEC adoption, the UK was an efficiency/effectiveness-maximiser.

66 Interview, London, 2 December 2009.

5 Vernetzte Operationsführung in Germany

The German variant of NCW is called Vernetzte Operationsführung (NetOpFü). The official English translation – although not quite resembling the German term – is Network-Enabled Capabilities (NEC). NEC was officially introduced into the Bundeswehr (the German Armed Forces) with the issuing of the *Conception of the Bundeswehr* in August 2004 (German Ministry of Defence 2004). In this document NEC was made the basis for an ambitious German military transformation project.[67] It took two years until – in 2006 – the first conceptual outline of NEC was issued (German Ministry of Defence 2006a). Conceptual work on NEC continued after 2006. Yet, the Bundeswehr struggled to implement NEC. Both an implementation strategy and a governance structure for NEC were missing. Early plans to have an initial operational capability (IOC) of NEC achieved by 2010 quietly disappeared from internal ministerial communications. In April 2009, the Chief of Federal Armed Forces Staff, in frustration over the length of time passed without tangible results, ordered that the NEC IOC needed to be achieved by 2012, and demonstrated in a military demonstration exercise (DemoEx) in spring 2013. Yet the exercise was cancelled in 2011 when it became apparent that necessary equipment would not be ready to use by 2013.

This chapter traces the NCW adoption process in Germany between 2001 and 2010. Similar to the British case study, it aims at recreating the processes by which the German military became aware of NCW, how the concept was adapted and how it was implemented. The ultimate aim is to assess the adoption pace and the patterns of NEC adoption in Germany and to link these variables to explanatory factors that are either efficiency/effectiveness-oriented or institutional legitimacy-oriented in nature. The results are discussed against the background of the theoretical framework in the concluding section of this chapter.

5.1 Introducing NEC into the German Armed Forces

5.1.1 Knowledge Acquirement

The introduction of NCW into the Bundeswehr took place amidst a profound restructuring of the German armed forces. In the beginning of the 21st century the Bundeswehr was still in a process of being adapted to a new mission spectrum. During the 1990s a number of events – the fall of the Berlin Wall 1989, the re-unification of Germany 1990, the integration of former National People's Army in the Bundeswehr, and the first out-of area missions of German soldiers – had rendered the structures, the equipment, the mission the Cold-War-Bundeswehr obsolete (Bredow 2005, 2007; Dyson 2005, 2008; Kümmel 2003; Longhurst 2005a, b; Longhurst/Miskimmon 2007). It is fair to say that Germany's armed forces were presumably affected the most by the changed strategic environment after the end of the Cold War, as its force structure had been entirely focused on (troop-intense) territorial defence. Between the years 1992 and

67 Formally, *Bundeswehr* and *German Armed Forces* have two slightly different meanings. The Armed Forces comprise only of the military branches whereas the term Bundeswehr includes not only the armed services but also the civilian employees. Both terms are often used interchangeably.

2000 international treaties as well as domestic political decisions required massive troop reductions and a number of military reforms.

Figure 1: Army Structures (Heer) and Overall Force Size since Re-Unification

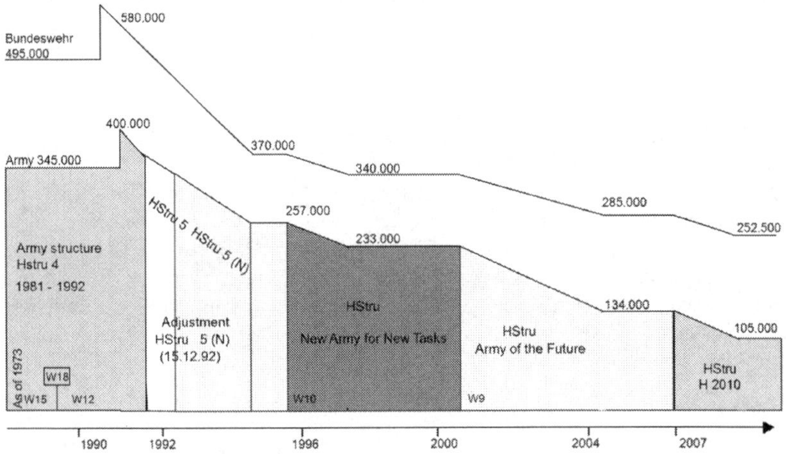

Taken from Klos et. al (forthcoming) BMVg Fü H).
Duration of obligatory military service in month (W18, W15, etc.).

In the result, the Bundeswehr and the single services had seen such a multitude of reforms, new structures (see figure above) and new distributions of competencies that one observer could not help but to label this constant adaptation as "cascades of reforms" (Bredow 2007). The German NEC project was tremendously intertwined with one of the more recent reform attempts, the *Transformation of the Bundeswehr* (2004–2010).[68] Even though it took rather long for the Bundeswehr to acknowledge the U.S. NCW concept in the first place, for a period of more than six years the German NEC received high attention in the MoD and was declared a *pillar* of German military transformation.

The *Defence Policy Guidelines* of June 2003 had opened a new era for the Bundeswehr by stating that the most likely mission for the Bundeswehr was stability operations rather than territorial defence. Even though NEC, NCW and joint interoperability were not mentioned in the 2003 Defence Policy Guidelines the document nevertheless paved the way for the transformation yet to come. When Defence Minister Struck tasked the Chief of Federal Armed Forces Staff Wolfgang Schneiderhan with producing a new *Conception of the Bundeswehr* in October 2003, he requested clarification on the following points: conscription, personnel reduction, re-structuring, operation planning, equipment and military capabilities. With regards to the latter, the Ministerial directive stated that, "new developments (i.e. NEC) have to be taken into account" (German Ministry of Defence 2003). NCW had thus reached the awareness of a critical mass of military and political decision-makers between summer and autumn 2003. By the end of 2003, NCW was discussed in the German defence community in a small number of publiccations and speeches (Scherz 2003; Wehrtechnik 2003).

68 In 2011 the Bundeswehr entered a new reform project, the *Re-orientation of the Bundeswehr* (Bundesministerium der Verteidigung 2011).

The adoption of NEC paralleled the introduction of the principle of *jointness* into the German Armed Forces. Before the issuing of the Conception of the Bundeswehr, the individual services were rather focused on maintaining interoperability with similar services within the framework of NATO and had no joint history in operations. Consequently, before the NEC project started a centralised institutional structure within the Ministry of Defence that would have ordered, sponsored or overseen information and knowledge acquirement on foreign joint operational concepts was simply missing. As a result, NCW knowledge acquirement in Germany was institutionally dispersed, undirected and dependent on entrepreneurship. Two main entry points for NCW knowledge acquirement can be identified: the Air Force Development Centre (Zentrum für Weiterentwicklung der Luftwaffe) and the Bundeswehr Centre for Analyses and Studies (Zentrum für Analysen und Studien der Bundeswehr, ZfASBw, later to become the Bundeswehr Transformation Centre). Other earlier potential opportunities to learn more about NCW (liaison and exchange), however, were passed upon.

5.1.1.1 The Limited Role of Liaison and Exchange Officers and Expert Groups

Some form of institutionalised military exchange between Germany and the U.S. existed when NCW appeared in the U.S. For example, in the year 2000, some 800 German Air Force officers were trained in Air Force bases Goodyear, Sheppard, Holloman, Pensacola Fort Rucker and Fort Bliss (International Institute for Strategic Studies: 2001). Flight training was rather practice-oriented, however, and did not involve any discussions of possible future command and control concepts.[69] Other institutionalised fora for potential idea-exchange included the Army Cooperation Staff (*Heereshauptverbindungsstab*) within the U.S. HQ TRADOC and the Air Force bilateral airmen-to-airmen talks. Yet, none of these channels had been used by the U.S. forces to push NCW or by the German individual services to request knowledge on NCW prior to the decision in 2004 to introduce the NCW concept into the German Bundeswehr.[70]

NCW knowledge acquired by liaison officers remained unused as well. A German Navy Captain, later to become German liaison officer to USJFCOM between 2003 and 2006, spent an academic year at the U.S. Naval War College in 1999. The College's president at that time was Arthur Cebrowski, the father of NCW. Reportedly, the Captain listened to three lectures given by Cebrowski on the subject of NCW,[71] yet he was not tasked by the Ministry to act as a multiplier when he returned to Germany.

Between 2001 and 2003 the U.S. NCW initiative was also known to military academics working at the German Armed Forces Command and Staff College (FüAk, Hamburg)[72] A joint publication by U.S. American and European think tanks on *Coalition Military Operations* was issued and one chapter in this book was devoted to NCW (U.S.-Crest et al. 2000). Yet, the importance of this publication was at best limited – a scientific attempt to discuss issues of multinational interoperability, "done by scientists, read by scientists"[73] that was not acknowledged by the MoD.

---

69 Interviews, Berlin, 30 January 2009 and Cologne, 13 March 2009.
70 Interviews, Berlin, 6 November 2008 and Cologne, 13 March 2009.
71 Interview, Strausberg, 17 March 2009.
72 Interview, Cologne, 13 March 2009.
73 Interview, Cologne, 13 March 2009.

Also the wider German military community did not pay attention to NCW when it was discussed in the U.S. either. This is evidenced by the fact that none of the widely-read German military journals (*Soldat & Technik, Europäische Sicherheit*) did publish a single article on Network-Centric Warfare before the year 2003. *Military Technology* – a publication in English – was the first journal to acknowledge NCW. It then was no other than Arthur Cebrowski himself who gave an overview of the basic tenets of NCW (Cebrowski 2003a).

Between 1999 and 2003 there was no significant push to deliver knowledge on NCW to Germany by the U.S. The U.S. Department of Defense stressed the need of its NATO allies to close the capability-gap. This pressure, however, was exerted on a more general level and rather with a view towards increasing defence spending rather than on specific conceptual programmes. On the other hand, there also was no specific interest from the German MoD to acquire knowledge on NCW. Even though the Armed Forces Staff within the Ministry tasked an expert group at the Joint Command and Staff College with looking out for new military trends, especially from the U.S., and even though a working group on 'operative planning and doctrine' was established in the late 1990s, information collection on NCW did not gather momentum.[74]

5.1.1.2 NCW Entry Points: The Air Force

The German Air Force played a key role in the NEC knowledge acquirement process. The Air Force Development Center (Zentrum für Weiterentwicklung der Luftwaffe, ZWELw) was a small think tank within the German Air Force in which some interest and expertise on NCW was concentrated from 2002 onwards. On behalf of the ZWELw, a seminar was conducted in 2003 to assess and evaluate the NCW concept (Marahrens 2006). Based on the findings presented to the Air Force Command, the Air Force Staff in the Ministry then issued a position paper on NCW in December 2003 (Chief of Staff of the German Air Force 2003). To compare, the Army Staff did not issue its position paper until June 2004 (Chief of Staff of the German Army 2004). Whereas this might not seem to be a big difference in time it was a first indicator for the general reluctance of the German Army to embrace NCW.

Relatively early in 2003, the ZWELw started spreading knowledge on NCW to the other services, arguing in favour of the concept for the Bundeswehr. It conducted joint seminars and workshops on NCW to keep track of international developments in this realm. One such notable workshop was the 'NCO Short Course' in September 2004 in which John Garstka, a key-figure for NCW in the U.S., participated (Marahrens 2006). The ZWELw also issued the first service publication on NCW (Schäfer 2005) and sponsored the translation of the CCRP publication 'Power to the Edge' into German (Alberts/Hayes 2006 [2003]). Furthermore, it developed the computer simulation CAFFEINE (Collaborative Game for first Experiences in a Networked Environment) that visualised the added value of NCW.

These first initiatives of the ZWELw, however, did not automatically lead to a defence-wide support for the NCW concept. Quite the contrary, Army officers argued that the Air Force was in 'cloud-cuckoo-land' when they believed in the promise of information superiority.[75] Furthermore, German history had resulted in a complicated rela-

74 Interview, Cologne, 13 March 2009.
75 Interview, Hamburg, 16 August 2009.

tion even of German military officers to physical violence. Many of the early (power point) presentations on NCW were focused on close enemy contact and were often based on combat scenarios. One U.S. participant in the 2004 Short Course, for example, proudly presented a video in which enemy forces were killed by U.S. troops. Instead of the expected applause the video presentation reportedly triggered an embarassed silence among the German officers in the audience.[76] NCW was thus criticised as not being suited for the kind of constitutionally approved crisis management missions the Bundeswehr was conducting.

5.1.1.3 NCW Entry Points: Bundeswehr Centre for Analyses and Studies

Apart from the German Air Force the Bundeswehr Centre for Analyses and Studies (Zentrum für Analysen und Studien der Bundeswehr, Waldbröl) was another entry point for NCW. The Centre was tasked by the MoD with looking out for new military trends in the late 1990s. For many years, the Centre only played a minor role in Bundeswehr modernisation and reform processes. It hosted an operational research core to produce net-assessment analyses, but such knowledge was less required in the post-Cold War era. However, in 1998 it was tasked with leading a long-term study on the future challenges for the Bundeswehr and the required military capabilities, with the results being issued in summer 2002. More than 350 individuals (from the armed forces, social and natural sciences and industry) contributed. The voluminous study covered international and societal developments, industrial developments, technological trends, key military technologies, possible developments of military organisations and armed forces and provided options for the improvement or generation of future required military capabilities. Although the report had also aimed at exploring current U.S. military changes, no explicit reference was made to network-centric concepts and their possible use for future Bundeswehr military doctrine.

In 2001, Col. Ralph Thiele assumed the post as the Centre's commander. Thiele, an Air Force officer, was an active supporter of the ideas of the Revolution in Military Affairs (RMA) and NCW. Right from the beginning, he tried to spread knowledge on NCW within the Ministry of Defence. In 2002 the Centre was tasked by the Armed Forces Staff with checking the current status of the U.S. RMA and assess if there was more to it.

Inspired by the British example of successful private-military collaboration, the Centre's working procedures were adapted. Three positions for industrial research fellows within the Centre were created with the aim of achieving a greater military-industrial exchange. Furthermore, the Centre created three Battle Labs and increased its number of German exchange officers in the USJFCOM from one to three. As a result, the USJFCOM invited Germany to participate in the Multinational Experiments Series MNE, which was concerned with developing interoperable C2-procedures for multinational coalitions. After the participation in the first MNE in 2001, the Chief of the Armed Forces Staff decided that the Bundeswehr should form a standing unit for Concept Development and Experimentation (CD&E) – which later on became involved in utilizing the German NEC. One of the German foreign exchange officers to USJFCOM at that time, concluded: “One can claim with justification that participation in the MNE

76 Interview, Hamburg, 16 August 2009.

series since 2001 was the inspiration for setting-up and developing a national CD&E capability for the Bundeswehr" (Neureuther 2008: 142, my translation).

In the year 2002, Thiele made several visits to the U.S. to meet individuals from the Pentagon, industry and think tanks to discuss future military trends. Personal ties existed with Andrew Marshall (the prominent figure in the U.S. RMA debate), Arthur Cebrowski, the Command and Control Research Program and the RAND Corporation.

In July 2004 (that is, prior to the issuing of the Conceptions paper in August 2004) the mission of the Centre – now re-named *Bundeswehr Transformation Centre* (Zentrum für Transformation der Bundeswehr, ZTransfBw) – was broadened. Modelled after the USJFCOM/J9, three new branches were established within the Centre: Concept Development & Experimentation, Modelling & Simulation and Capability Assessment.

In sum, the Transformation Centre served as an entry point for the NCW concept from 2002 onwards.[77] In a spiral process, the Armed Forces Staff in the Ministry of Defence was informed by the Centre about NCW and, in return, tasked the Center with gathering more information on the concept. Through the increased appreciation of the Centre by the Armed Forces Staff, the Centre improved its visibility and strengthened ties with the USJFCOM.

5.1.1.4 The Minister of Defence and the Chief of Federal Armed Forces Staff

The Transformation Centre also initiated the first publication on the concept of NCW and its possible value for the German Armed Forces. The book, *Vernetzt zum Erfolg* (Networked for Success), published in the summer of 2003, contained a state-of-the-art description of the NCW concept and a discussion of NCW's potential values for the Bundeswehr (Mey/Krüger 2003). Most importantly, the Chief of Federal Armed Forces Staff, Gen. Schneiderhan, contributed the foreword to the volume, wishing that the book "would gain a high circulation within the Bundeswehr, the military industry and the interested public" (my translation). In November 2003 Schneiderhan gave a speech before the Federal Association of the German Industry (Bundesverband der Deutschen Industrie) on network-enabled capabilities.[78] This speech, held one month after the Defence Minister Struck directed Schneiderhan to work on a new Conception of the Bundeswehr, was the first official explanation of the German position on the issues of NCW and transformation and their added value for German Defence.

The Chief of Federal Armed Forces Staff was a central figure in the process of NCW knowledge acquisition. He showed an open attitude and active interest on the various technical and non-technical aspects of the U.S. military transformation. In the year 2003 Schneiderhan, together with the Chiefs of Service Staffs, paid several visits to the Pentagon (especially the Office for Force Transformation), USJFCOM and NATO's Allied Command Transformation in order to discuss the basic tenets of the current U.S. transformation.[79] After meeting the Commander of the USJFCOM and Commander of the ACT, General Giambastiani, Schneiderhan reportedly said that the American kind of

77 Military officials interviewed for this book stressed the role of Thiele and the Bundeswehr Transformation Centre in the knowledge acquirement process and spreading knowledge about the concept within the German military community.

78 Accessible online at www.bmvg.de.

79 Interview, Strausberg, 17 March 2009.

transformation was exactly what the German Armed Forces needed.[80] A further important meeting took place in December 2003 at the National Defence University with Chief of Armed Forces Staff Schneiderhan and Arthur Cebrowski, after which Schneiderhan – briefed about the U.S. military transformation – said that Germany was in need of a military transformation as well.

In general, the authority of a Chief of Federal Armed Forces Staff is limited. Still, their influence can vary along the political-military continuum. Whereas, for example, former Chief of Federal Armed Forces Staff General Klaus Naumann most likely will be remembered as a rather political figure that led the debate on the participation of the Bundeswehr in post-Cold War military missions, Gen. Schneiderhan was less concerned with international security issues but focused on improving mission-readiness of the Bundeswehr. He not only endured the transition from Defence Ministers Scharping to Struck (both SPD), but also received the confidence of Defence Ministers Franz-Josef Jung (2006–2009, CDU) and Karl Theodor zu Guttenberg (2009–2011, CSU). Schneiderhan's official retirement date was extended until he was dismissed by zu Guttenberg in November 2009 over the so-called Kunduz affair.

The relation between Schneiderhan and Defence Minister Struck has been described as being constructive and characterised by a mutual commitment to make the Bundeswehr fit the new operational challenges, especially in Afghanistan. Defence Minister Struck also supported the capability-driven version of the military transformation that evolved in the U.S. When Struck met U.S. Secretary of Defense Rumsfeld on the occasion of the annual Munich Security Conference in February 2004, Rumsfeld reportedly took a piece of paper to draw on it the basic principles of U.S. transformation. Struck took it, laughing, and said that was exactly what he was going to present to his ministry (SPIEGEL 2004).

5.1.1.5 NATO Transformation

In German military circles the terms *transformation* and *network-centric* gained prominence only after the set-up of NATO's Allied Command Transformation (ACT) and the issuing of the NATO transformation agenda.[81] The transformation of NATO and the development of a NATO approach to NCW played a decisive role in the knowledge acquisition process. At the Prague Summit in October 2002, NATO leaders committed themselves to military transformation, set-up a NATO response force and the Allied Command Transformation (ACT), which was formally established in June 2003. Two German Liaison Officers represented Germany in ACT and up to 60 seconded German officers served in ACT. In the beginning, German officers were mainly Navy staff, as ACT formerly hosted the Allied Command Atlantic.

In July 2003, a high-level NATO Seminar was held on the observations drawn from the U.S. and UK operational experiences in Iraq. The seminar results identified NCW as one important capability NATO forces should obtain in the future. In this context, the two NATO Strategic Commands developed a *Strategic Vision* that stressed the importance of the effects-based operations approach to which NCW was crucial (SHAPE and ACT 2004). At that time, NATO had already initiated a study on NCW (U.S.) and the British NEC concept to explore if and how NCW could be of importance for NATO

80 Interview, Berlin, 6 November 2008.

81 Interview, Glücksburg, 9 February 2009.

(NATO Consultation Command and Control Agency 2005). Germany was then one of twelve NATO members to fund the study. Later, the dedication to military transformation and NCW became institutionalised for Germany within the ACT, and by 2005 the German fraction in ACT was one of the most active to foster NATO transformation.[82]

The set-up of ACT could be understood as an attempt by the U.S. administration to force European allies to hold up to their capability commitments. The U.S. filled the most strategic posts in the Alliance: Supreme Allied Commander Europe (SACEUR) and the Supreme Allied Commander Transformation (SACT). Until 2009, the Commander to ACT war also the Commander of the USJFCOM, which was in charge of implementing U.S. military transformation (through experimentation, joint training, joint capabilities development and as a force provider).[83]

In the years 2002 and 2003 the U.S. reportedly threatened behind closed door to de-emphasise their engagements to NATO if the ACT did not become a success and if the Allies did not transform their militaries in order to become interoperable and share the military burdens of global engagements.[84] Analogous to Albert O. Hirschman's concept, in which he identified exit, voice and loyalty as potential strategies of members of an organisation to respond to organisational decline, the U.S. used the "threat of exit"[85] to achieve the compliance of the other NATO members (Hirschman 1970). At the same time, the ACT also managed to spread enthusiasm on the NCW issue. The first Commander of ACT, Admiral Edmund Giambastiani, who was previously the Senior Military Assistant to Rumsfeld in the U.S. DoD, was a passionate proponent of Rumsfeld's techno-centric transformation. Due to his strong personality and his dedication to the transformation process, he established a specific pro-transformational mind-set among the first foreign officers to ACT.[86] The 60 German officers on secondment kept close ties to the Armed Forces Staff, thereby spreading knowledge and enthusiasm of NCW to Germany.

5.1.1.6 Conclusion

Knowledge acquisition started rather late in 2003 and progressed throughout 2003 and 2004. The process was largely decentralised and benefited from devoted entrepreneurs at the senior level that actively tried to foster awareness of NCW within the Ministry and the individual services.

For a long time, there was no institutional awareness within the Ministry of Defence, especially in the Armed Forces Staff, of the potential importance of NCW. Reports by liaison officers in the U.S. on NCW did not result in increased information gathering. This could be explained by a number of factors. First, between 2000 and 2002, the Bundeswehr was in the middle of a difficult reform process, yet this reform was not so much directed at increasing operational capabilities but towards streamlining adminis-

82 Interview, Berlin, 12 November 2008.

83 In 2009 France rejoined the NATO Command Structure. As a result, the distribution of high-profile posts in NATO changed and French General Stéphane Abrial assumed command.

84 Interviews, Berlin, 6 and 12 November 2008, and Glücksburg, 9 February 2009. Frans Osinga made a similar point (2010: 30). Heiko Borchert, however, argues that German politicians did not believe in the wholeheartedness of the threat (Borchert 2010: 107).

85 "The threat of exit (...) [is] an instrument of voice" (Hirschman 1970: 86).

86 Interview, 9 February, 9 February 2009.

trative and business procedures. Furthermore, there was neither an Armed Forces-wide digitisation programme already in place nor an initiative to increase jointness in general that could have paved the way for NCW. A visible push from the defence industry was also missing. As a result, there were only minor attempts to gather information on C2 issues that, in consequence, did not have much impact. In sum, during the first years after the development of U.S. NCW there was only little breeding ground for the concept inside the German Bundeswehr.

Table 11a: The Timing of NEC Knowledge Acquirement in Germany

Timing of Knowledge Acquisition	Moderate/Late
Explanatory Factors for Delay	**Nature of the Factor**
Ongoing Difficult Reform Process	Institutional Legitimacy
No Strong Joint Principles, no Aim of Joint C2 Structures	Institutional Legitimacy, Military Culture
Explanatory Factors for Knowledge Acquisition	**Nature of the Factor**
Early Entrepreneurship (Working Level)	Agency
Moderate U.S. Push/Information Exchange	Institutional Legitimacy

Eventually, however, interest for NCW increased. This was initially due to the entrepreneurship of the Air Force Development Centre and the Bundeswehr Centre for Study and Analysis (later Transformation Centre). Furthermore, invitations and presentations by the USJFCOM and the OFT increased awareness that something relevant was going on in the U.S. The role of the U.S. in the process was twofold. At the political level, it exerted pressure on Germany to modernise its military capabilities threatening to exit NATO. At the working level the US provided information on the military transformation (through briefing and seminars), invited foreign exchange officers to USJFCOM and enabled German participation in the MNE series. Until it was decided to introduce NCW, the pace of knowledge acquisition could be described as moderate at best, as it took more than one year to announce NEC in the Conception of the Bundeswehr. This was due to an internal opposition mainly coming from the German Army, as NCW was seen as a combat-oriented concept unsuitable for crisis management missions.

Apart from this rather functional argument against NCW and thus, an own German NEC project, a more fundamental debate evolved in 2003 about the merits (or, drawbacks) that NEC could have in the area of military leadership. For some critics NCW jeopardised one core principle of German military culture – mission command. Mission command is "a philosophy of command that requires and facilitates initiative in all levels of command directly involved with events on the battlefield. It allows and encourages subordinates to exploit opportunities by empowering them to demonstrate initiative and exercise personal judgment in pursuance of their mission while maintaining alignment through the commander's intent." (Shamir 2008)

In the German debate, arguments were made that through the networking across military hierarchies the command style could unintentionally be transformed to favour micromanagement by the superior rather than initiative by the staff (Lange 2008; Laupert 2008; Mey/Krüger 2003). The image of the Defence Minister sitting in front of a plasma screen directing forces on a battlefield thousands of kilometres away was evoked by opponents of NCW theory as a negative example of what NCW could mean for the German armed forces years before the iconic picture of U.S. President Barak Obama monitoring in the White House Situation Room the CIA-led operation to capture Osama bin Laden travelled around the world.

A final factor that slowed down knowledge awareness was the decentralised institutional structure. Technical aspects of NCW raised the attention of the Armaments Department in the Ministry; while the Centre for Analyses and Studies was attentive to the conceptual underpinnings of NCW. The individual services had their own air-, sea- or land-based visions of the concept. The Armed Forces Staff tried to act as a broker between all parties but there was no concerted action.[87]

Nevertheless, knowledge acquisition in Germany gathered steam eventually. The NATO transformation agenda and the beginning of the NATO Network-Enabled Capability project raised a general awareness for C4ISR-issues. This was paralleled by a sudden interest from the Chief of Federal Armed Forces Staff Schneiderhan and the Defence Minister Struck for a military transformation modelled after U.S. transformational concepts. *Transformation*, *network-centricity* and *effects-based operations* – all of these new concepts sounded intriguing as they offered a rhetorical escape from the deadlocked reform legacy of the past.

Table 11b: The Pace of NEC Knowledge Acquirement in Germany

Pace of Knowledge Acquisition	Moderate/Slow
Explanatory Factor	Nature of the Factor
Critical of Combat-Oriented Military Operational Concepts	Organisational Culture
Scepticism towards Impact on Mission Command Principle	Organisational Culture
No Dedicated Institutional Entity	Institutional Legitimacy/Path Dependency
NATO Push	Institutional Legitimacy
Late Entrepreneurship (Politico-Military Level)	Agency
Lookout for Transformational Concepts	Institutional Legitimacy

In a nutshell, the NCW knowledge acquisition process was characterised by strong mid- and top-level entrepreneurship, a domestic window of opportunity for a new reform, the availability of NCW as a ready-made concept which value had already been proven, the interest of the U.S. in spreading the concept and the push from NATO in this direction, including the more general political pressure to modernise German military capabilities. It was especially the mid- and top-military elites that actively requested knowledge on NCW. One military official interviewed for this book concluded that, "foreign exchange and liaison officers, participants in multinational fora, conceptual transformationalists but also high-ranking decision-makers attending official consultations and visits – these were the 'trend scouts' for the Bundeswehr, or, the entrance door for gaining knowledge".[88] Operations commanders – practitioners in the theatre – were apparently not involved. Especially in the beginning, NEC was considered a purely technical matter best dealt with by IT specialists. It was presumably not perceived as a career opportunity and thus less attractive for officers.

The knowledge acquisition process was also characterised by its decentralised institutional structure, as there was no central institutional entity inside the MoD that acted

87 Interview, Hamburg, 26 August 2009.
88 Interview, Berlin, 30 April 2009.

as an entry point and that, at the same time, had enough authority to spread information about NEC. Instead, there were at least two rather peripheral institutional units that were relevant for early knowledge acquirement on NCW: The Air Force Developments Centre and the Centre for Analyses and Studies that tried to gain the privilege of interpretation over the NCW issue. The next section will show that this decentralised structure also had an impact on the slow utilisation of NCW.

5.1.2 Utilising NCW as NEC in Germany

5.1.2.1 Introducing NEC to the German Defence Community

The German NEC was officially acknowledged for the first time in the Conception of the Bundeswehr (9 August 2004), which was written by the Armed Forces Staff Directorate VI Referat 2 (FüS VI 2). The Conception of the Bundeswehr outlined a profound military transformation of the Bundeswehr. The aim was to "improve the mission readiness of the Bundeswehr. To achieve this aim, the tasks, capabilities, equipment and financial resources have to be synchronised in an overall and joint approach" (German Ministry of Defence 2004).

NCW was considered important for this transformation, as "achieving this ability is a core element of the transformation and to be promoted with priority" (German Ministry of Defence 2004: 13). NEC was defined as the "command and deployment of armed forces on the basis of a joint interoperable Bundeswehr communication and information system across all command levels, which connects all relevant individuals, locations, units, facilities and sensors and effectors with one another" (English translation by Collmer 2007; German original in German Ministry of Defence 2004: 12). Yet, apart from this rather superficial definition, there was neither an outline of the tenets of NEC nor a vision on how NEC would contribute to the Bundeswehr mission profile. In fact, NEC had not even been conceptualised when it was made the core feature of German military transformation.

The decision to implement NEC was taken by the Chief of Federal Armed Forces Staff only weeks before the Conception of the Bundeswehr was issued in summer 2004.[89] Prior to this decision, the services, but also other directorates (IT and Armaments) provided position papers on NEC: The Air Force as early as December 2003 (Chief of Staff of the German Air Force 2003), the Army only in June 2004 (Chief of Staff of the German Army 2004). Shortly before the issuing of the Conception of the Bundeswehr in July 2004, a Transformation Coordination Group was set-up that had decision-making competencies within the MoD.

5.1.2.2 Drafting the German NEC Concept

Compared to other transformational concept papers, the drafting of the NEC concept took a considerably long time. The sub-divisional document on NEC, the 'TK Net OpFü', needed more than two years to be finalised. In the Conception of the Bundeswehr the issuing of 38 such sub-divisional documents was announced (for example on Combat Search and Rescue, Computer Network Operations, and Target and Battle Damage Assessment). Many of these were published throughout 2005. In the case of

89 Interview, Berlin, 6 November 2008.

NEC, however, the Armed Forces Staff only issued the NEC sub-divisional paper in November 2006. The opposing visions of single services about NEC had hampered the finalising of the document. Over the years, the ministry and the services had developed a corporatist system of joint underwriting (Mitzeichnung) in which every major decision – such as the NEC sub-divisional paper – had to be approved by the service staffs and the civilian directorates. The composition of the NEC working group, furthermore, did not suit the actual task of contributing to a doctrinal instead of a technical document. This problem is captured in the following quote: "The services all sent IT-personnel from their command support commands. Yet what we would have needed was personnel from the operations commands. But the services still thought NEC was only about IT."[90] Even in 2009 all NEC-delegates of the services and the Armaments Directorate except for the Army Staff and the Armed Forces Staff had a professional IT background.

Another reason for the slow utilisation was the "German culture of X-raying everything in advance in order to assess potential weaknesses".[91] The Army insisted on clarifying its concerns on mission command procedures and was especially suspicious of the possibility that NEC would jeopardise existing command styles. As a result, the NEC sub-divisional paper remained formulaic and did not make a link to actual operational challenges.

In sum, the process of coming to a more detailed NEC definition took exceptionally long. Even though the Armed Forces Staff became the central institutional entity in utilising NCW, it did not manage to bring together the different understandings of NEC or to forcefully push a joint NEC concept. This problem did not cease to exist. In 2010 the Ministry announced a plan to revise the 2006 NEC sub-divisional paper.[92] Anticipating potential coordination problems, the envisaged date of issuing the revised document was set to 2013.

5.1.2.3 Concept Development and Experimentation

Research on NEC did not stop with the issuing of the NEC sub-divisional paper in 2004. Conceptual work continued in the framework of Concept Development and Experimentation (CD&E). CD&E is a method to verify military doctrines, management methods and technical innovations (Honekamp 2008). It was mainly developed in the U.S. and was a main tool by the USJFCOM to test transformational military concepts. While CD&E was not pursued rigorously by many countries, the Bundeswehr CD&E department was the second largest CD&E unit worldwide (only outnumbered by USJFCOM) modelled after the USJFCOM counterparts.

With regards to NEC, the Transformation Centre initiated and supervised the joint biennial national experiment series *Common Enhancement*. This was paralleled by service experiments (the Navy's sea basing project and the Army's NetOpFü series). Although the experiments were generally regarded as a success, several critiques have been raised on various issues. First, senior officers complained that investing in comparatively cheap experiments was just an excuse for not investing in the actual procure-

90 Interview, Bonn, 20 February 2009.

91 Interview, Bonn, 20 February 2009.

92 Presentation given at the NEC Course at the German Armed Forces Command and Staff College, Hamburg, 7 June 2010.

ment of C4ISR equipment. Even though the experiments received some attention in the defence community NEC's focus remained conceptual and directed towards soft or *heavy-headed* issues (notably understanding human cognitive capacities).[93] A further critique concerned the actual value of the experiments, as there was no ministerial directive on how experimentation *results* would be fed into the services. It remained with the individual services to explore the findings derived from joint experiments – or not. Finally, experiments were conducted by an ad-hoc assemblage of military contingents. Participants would only come together for the experiments and then return to the units they had been drawn from. Even though the Command and Staff College proposed the set-up of a standing NEC training brigade in 2008 (modelled to some extent after the U.S. Stryker brigades), the proposal did not receive the necessary support from the Army due to a lack of funding. As a result NEC competence was not exploited sufficiently.

In sum, CD&E was rhetorically coined as the method by which to *implement* NEC; yet, it appeared that through the prominence of the CD&E method the NEC project remained *conceptual.*

5.1.2.4 German NEC and NCW Compared

Turning to the substantive features of the NEC concept reveals that in contrast to the British NEC concept the German variant of NCW showed a high degree of concept faithfulness.[94] Although there were some variations between the two concepts especially with regards to the 'role of the individual' both concepts showed more similarities than differences.

In the NEC sub-divisional paper the authors borrowed extensively from the U.S. concept. Similarly to the American documents, the NEC sub-divisional paper linked NEC to a broader military transformation project. Conceptually, the German NEC also rested on the 4-Domain model and the NCW-value chain (simple version of the NCW RtC and not the more elaborate version presented in the NCO CF). Emphasis was given to *common situational awareness* and not so much to the overall mission effectiveness. Yet, despite this slightly different touch the overall aim of NEC – gaining *superiority* over an enemy – was repeated.

A major difference between NEC and NCW concerned the role of the individual. Although the NCW concept had progressed in the NCO CF to paying more attention to the human dimension of warfare[95] it was still too negligent of the human individual. In the German understanding of warfare a network could never be the centre of a military enterprise. *Der Mensch im Mittelpunkt* – the individual in the focus – was thus one of three transformational principles in the Conception of the Bundeswehr referred to also in the NEC sub-divisional paper. The consequence is one of nuance, but is nevertheless important. Whereas the human dimension in the NCO CF was concerned simply with

93 Interview, Berlin, 6 November 2008.

94 The comparison is based on the following documents: The U.S. DoD NCW Report to Congress (NCW RtC) of July 2001, the OFT/OASD-NII Network-Centric Operations Conceptual Framework Version 2.0. (NCO CF) (Garstka/Alberts 2004; U.S. Department of Defense 2001) and the NEC subdivisional paper (German Ministry of Defence 2006a).

95 The NCO CF touched upon the issues of teamwork, the sharing of knowledge, the challenge of a common understanding of an issue or the problem of bringing the right knowledge to the right people.

finding the right people to achieve NCO (Chapter: 'The Network Centric People') the German NEC discussed the problem of adjusting NEC command procedures to be user-optimised. This difference in approaching the individual in both NCW and NEC was also mirrored in the German term for NEC. *Vernetzte Operationsführung* did not carry the notion of centricity. *Vernetzte Operationsführung* could roughly be translated into 'the networked command of operations'. The network was thus only an attribute; the command of operations remained a human task.

The third observation on the differences between the U.S. model and NEC – and an issue of considerable debate within the German military community – concerned the principle of mission command. Some feared that NEC could undermine mission command. Others argued that on the contrary, NEC would enable subordinates in the field to become more independent from the command level and thus even increase mission order tactics (Klos 2006; Meiter 2006). In the end, the latter perspective prevailed and the NEC sub-divisional paper stated: "Through end-to-end networking, NEC enables a straight control across all command levels. Yet direct control has to be an exception and mission command remains the valid principle in force" (German Ministry of Defence 2006a: 18).

A related debate touched upon the issue of *empowering people*. The Army was critical of the organisational consequences of enabling military units through networking. It opposed the assumptions made by the authors of the volume *Power to the Edge* (Alberts/Hayes 2003), a book that was so highly valued by the German Air Force that had supported the translation of the volume into German. Apparently, empowering 'edge' people by giving them more leeway meant something different if one pilot represented the edge or a platoon with 50 infantrymen.

Finally, and in contrast to U.S. documents, the NEC sub-divisional paper introduced definitions for gradual levels of NEC-readiness. NEC should come to life, first, through achieving common situational awareness, second, through achieving integrated command procedures, and third, through achieving an integrated reconnaissance-command-engagement chain (German Ministry of Defence 2006a: 17). This represented a holistic, conceptual top-down approach to NEC implementation. The U.S. approach was more pragmatic and favoured the creation of mission-tailored capabilities packages. Using a similar pragmatic approach, the NEC sub-divisional paper could – for example – have requested the immediate development of beacon projects important for the forces deployed in Afghanistan such as Blue Force Tracking capabilities. Instead the NEC implementation was to be achieved rather evenly over a wide range of military units. This holistic conceptualisation of NEC implementation caused frustration in the military.

To conclude, the German NEC concept was philosophically considerably based upon the original U.S. concept. It introduced the domain model and the value chain, kept the notion of superiority over the enemy and stressed the value of NEC for achieving jointness. The concept, furthermore, also emphasised the method of CD&E to achieving NEC. On the other hand, NEC departed in various aspects from NCW/NCO by stressing both the role of the individual and the principle of mission-order tactics. The overall ambition of NEC appeared mixed. On the one hand, the German NEC conceptually aimed at achieving the *superiority* of effects. On the other hand short-term implementation targets were much humbler (achievement of a common operational picture and situational awareness). Yet the proposed way of implementing NEC was not pragmatic and being oriented on current operational scenarios but rather holistic and intellectual in nature.

5.1.2.5 Motivations for the Introduction of NEC in Germany

What motives and beliefs underpinned the introduction of NEC? In the Conception of the Bundeswehr *mission-orientation* had officially been presented as *the* rationale for the Bundeswehr transformation and, thus, for the adoption of NEC. And like a reflex, almost all military officers interviewed for this book pointed out an "improvement of the operational capacities" through NEC. It seemed, however, that mission-orientation was just a buzzword. These statements contradicted the fact that for years NEC remained conceptual and much man power was devoted to further detailing the approach than implementing it. Against the background of very different mission statements of the Bundeswehr and the U.S. armed forces it is also puzzling that the German NEC closely resembled NCW. Moreover, if mission orientation had been at play the exceptionally long time of conceptualisation was hard to explain. Finally, theatre assessments made by German defence experts revealed that NEC was not a priority for ISAF (Lange 2008). In sum, mission orientation might have rhetorically been the rationale for adopting NEC, but this was not mirrored in actual conceptual design.

No empirical evidence could be found that NEC had been introduced out of *national ambition*. Even though the Defence White Paper of 2006 mentioned the German "responsibility" to support the provision of global security and pointed at the interrelation between the current transformation and German security interests (German Ministry of Defence 2006b), national ambition, or prestige, did not steer the introduction of NEC. Also, no evidence could be found to support a neorealist argument (that is to increase German relative military power) for the introduction of NEC. For the case of NEC adoption the general conclusion on the German role in international security after the end of the Cold War offered by Sebastian Harnisch and Kerry Anne Longhurst seemed to apply: "(t)here was simply neither the financial resources nor the political will to expand Germany's international role" (Harnisch/Longhurst 2006: 54).

The same limited role applies to *industrial lobbying*. The defence industry did not exert a significant influence during the process of utilising NEC. This was mainly due to two reasons: First, there is a strict constitutional civil-military divide in the Bundeswehr and, in contrast to other countries, there is only little possibility of joint military-private collaboration in military research and technology. This constitutional delineation between the military and the industry prevented a close cooperation in NEC development. Second, in the beginning, the industry lacked interest in the subject. In 2002 the Transformation Centre, for example, created an industrial research fellows exchange opportunity. Yet, this opportunity was not of interest to the industry and research positions remained open. The Bundeswehr Centre for Analyses and Studies even proposed the establishment of a standing military-industrial research cell modelled upon the British NITEworks enterprise but also this collaboration effort failed. Apparently, NCW and NEC were less interesting for the German defence industry. The emphasis on C2 did not correspond with existing German industrial expertise (aviation, ballistic missiles and protected vehicles). The German procurement system, furthermore, was also more or less closed to foreign competitors so that, as a result, no foreign influence came to bear.

The *role of the U.S.* in the NEC introduction process was interesting. *Interoperability with U.S. combat forces* did not play a role. Neither has there been direct pressure from the U.S. to adopt NEC. The U.S. did, however, exert implicit pressure through the channel of NATO. References to the NATO NNEC process can be found in various German documents. The 2006 Defence White Paper, for example, justified the introduction of NEC with a reference to NATO: "In NATO and the EU, armed forces planning is oriented on the principles of NEC" (German Ministry of Defence 2006b). The NEC sub-divisional paper devoted a whole section on the NATO NNEC process and, even in the Conception of the Bundeswehr of 2004, NATO influence on German capability development was stressed.

A *domestic window of opportunity* was a further factor for the introduction of NEC. The U.S. transformation model and NCW as a ready-made and combat-proven concept appeared at a time when the German civilian and military leadership were in desperate need of a new rationale for what was considered a necessary new military reform attempt. In the beginning of the 2000s the incoherence between the Bundeswehr's structure and the operational reality that the troops faced – especially in Afghanistan – frustrated many in- and outside German defence. By the time the new Defence Minister Struck took office in 2002, many staff officers and soldiers were tired of supporting what would supposedly become the "reform of the reform" (Inacker 2002). *Transformation* and *NEC* were new and intriguing concepts that invoked a forward-looking perspective and held the promise of modern and 'trendy' equipment. It could be argued that the change from Minister Scharping to Minister Struck opened a *window of opportunity* for the introduction of a reform and has contributed to the tenacity with which the concept has been introduced into the German Armed Forces.

Increasing jointness and combinedness was another important factor for the introduction of NEC. In a sense, NEC was both the cause and consequence of a stronger emphasis on the joint principle. For many decades the German Armed Forces did not have strong joint features. The Ministry of Defence, naturally, provided joint overview of military affairs, but the armed forces had not created joint command structures. By the end of the Cold War the German Armed Forces were organisationally and technically rather linked to similar services of other NATO countries than to their own national complements. Organisationally, this was expressed by the absence of a joint operational command. As national and allied defence was the sole scenario for combat action and would, moreover, result in a common NATO response, there was no rationale for setting-up a national joint command during the Cold War.

This changed during the 1990s when the mission spectrum of the Bundeswehr was broadened. Although the need for reforming the operational command and control structures had been recognised as early as August 1990, an adequate political initiative failed to appear. Due to political opposition and the parochial interests of the individual services that intended to keep their own command structures, the issue remained largely unresolved during the 1990s (Young 1996). The establishment of joint command structures only succeeded in 2001 with the set-up of the Bundeswehr Operations Command. Interestingly, MoD staff interviewed for this book named *jointness* quite often as a motive for NEC. At the service level, however, officers still understood NEC as a means to increase *combinedness* with respective NATO partners.

5.1.2.6 Conclusion

The tracing of the NEC-utilisation process and the comparison of NEC and NCW provided insights on the differences between the two concepts. Moreover, this sub-section also focused on the factors that accounted for these differences. Both the pace of NEC-utilisation and the assessment of the degree of concept faithfulness allow for making assumptions concerning efficiency/effectiveness and institutional legitimacy as drivers for the German NEC programme.

The pace of utilising NCW and conceptualising NEC in Germany was rather slow. It took more than two years to draft the sub-divisional paper on NEC in 2006 and an announcement in 2010 to revise the NEC-Strategy by 2013 re-affirms this conclusion. This slow pace was startling against the background of continued presence of German military personnel in Afghanistan and the urgent need for adequate operational concepts.

There are a number of factors that could explain the slow pace of NEC utilisation. First, strong military leadership was missing. Instead, the consensus principle and the practice of joint underwriting hampered a pragmatic formulation of the concept. The absence of a strong institutionalised military leadership and the weak discretionary power of the Chief of Federal Armed Forces Staff had their origins in the decision to not set-up a general staff when the Bundeswehr was established in 1955 (Young 1994: 12). Only in 2005 the competencies of the Chief of Federal Armed Forces Staff had been gradually increased (German Ministry of Defence 2005).[96]

Table 12a: The Pace of NEC Utilisation in Germany

Pace of Utilisation	Slow
Explanatory Factor	**Nature of the Factor**
Lack of Centralised Authority/Consensus Principle/Joint Underwriting	Institutional Legitimacy
Lack of Support for NCW	Institutional Legitimacy
Conceptual Approach	Organisational Culture

Between 2000 and 2005 The Bundeswehr Chief of the Armed Forces Staff and the Armed Forces Staff did not have the necessary authority over the other services, let alone the civilian directorates in the Ministry to direct NEC conceptualisation. As a result, like other joint or defence-wide projects also NEC was subject to long-winded negotiations between the civilian departments (procurement and modernisation), the individual services and the joint military departments.

There was also a lack of support for NEC from within the individual services. Even though NCW had have sponsors in the concept development branches of the services others in the Navy and the Air Force argued 'network-centricity' had already been achieved and that, from an operational perspective, they would rather seek continued integration with their respective NATO commands than jointness with the German Army. The Army, on the other hand, was likely to carry the main burden of digitisation and was therefore keen on reducing the potential impact of NEC on its structure and procurement projects. More generally, there was concern about the offensive nature of NCW – especially at the time of U.S. military engagement in Iraq to which the German political elite was critical of.

Finally, German military culture slowed down the utilisation of NCW for the Bundeswehr (Wiesner 2010). There has been a culturally grounded dislike for 'quick' solutions and risk-taking in German culture (Hughes/Werwatz 2006). This has been cap-

96 In the wake of the *Re-orientation of the Bundeswehr* the role of the Chief of Federal Armed Forces Staff again was strenghtened. In 2012 the position received competencies almost comparable to joint chiefs of staff in other countries (German Ministry of Defence 2012).

tured in the so-called *uncertainty avoidance index* (UAI), which was used in a study on values and culture in the field of international business management.[97] The UAI "deals with a society's tolerance for uncertainty and ambiguity (...). It indicates to what extent a culture programs its members to feel either uncomfortable or comfortable in unstructured situations. (...) Uncertainty avoiding cultures try to minimize the possibility of such situations by strict laws and rules, safety and security measures, and on the philosophical and religious level by a belief in absolute Truth; 'there can only be one Truth and we have it'".

In a comparative research project it was established that the German UAI was nearly twice as high (65) than the British (35). If one agrees that the military as part of the national societal system is influenced by national cultural variables then a general risk averseness can be assumed at least for the bureaucratic branches in the Bundeswehr, too. It appears that in the case of NEC utilization the Bundeswehr focused on time-consuming concept-development to get to the heart of NEC rather than on finding solutions for operational problems in ISAF. Heiko Borchert made a similar observation (Borchert 2010: 86): "Germany's transformation process is very much concept driven. Other countries follow a more pragmatic and hands-on transformation approach that aims for quick wins. The Bundeswehr, like the German public sector in general, by way of history and cultural heritage, has a tradition for serious conceptual work and sound deliberation before actions are taken. This inclination dates back to Max Weber and his analysis of the inner workings of bureaucracies. The bureaucratic culture in Germany also explains why it is so important to get new conceptual ideas into the paper trail before things get done. This is noteworthy because the ideas of transformation and [NEC] entered the new Bundeswehr concept papers relatively late".

Concept faithfulness, the second variable discussed here, was rather high in the case of NCW utilisation. The German NEC concept shared the basic tenets of the U.S. version. It differed only in that it rejected the notion of network-centricity and put the individual into focus instead. Yet, compared to the British techno-scepticism, the German emphasis on the individual was not so much a question of training or enablement but one of accountability, responsibility and morale. A further delineation from the original U.S. concept was the intellectual, formulaic style in which the concept was presented. Even though the building blocks were similar to the original U.S. version, the German version appeared to be detached from actual military scenarios and operational reality and remained rather theoretical. The table below summarizes the factors that can account for the high concept faithfulness and for the only minor variation of the concept.

97 See Geert Hofstede's project's website: http://www.geert-hofstede.com (accessed 13 July 2010). Also see Hofstede and Hofstede (2005).

Table 12b: German NEC Concept Faithfulness

Concept Faithfulness	High
Explanatory Factor	**Nature of the Factor**
Lack of Military Strategy and Doctrine	Institutional Legitimacy
Concept Variation 1: "The Individual into the Focus"	
Explanatory Factor	**Nature of the Factor**
Responsibility and Accountability	Military Culture (Institutional Legitimacy)
Concept Variation 2: Intellectual Approach and Formulaic Style	
Explanatory Factor	**Nature of the Factor**
Decoupling of NEC from Operational Pressure	Institutional Legitimacy
Conceptual Approach	Organisational Culture

The similarities between the original U.S. version and the German concept were puzzling against the background of different force postures of the two armies during the last decade. Whereas the U.S. military was engaged in the 'Global War on Terror' and conducted high-intensity and special operations during the Iraq invasion and in Afghanistan, the German mission reality was stabilisation rather than peace enforcement. Yet the NEC-concept was not adapted so as to mirror this reality. A national military strategy and doctrine, which would have guided NEC adaptation was missing. The German political elite was reluctant to clearly define German strategic interests and to derive conclusions for adequate force structuring and military doctrine. This inability to clearly articulate the link between actual military operations and NEC was a result of the constraints from the political level and, thus, based on institutional legitimacy.

The stressing of the individual within the concept represents one variation of the concept. However, the focus was not on *enabling* the individual, which would have implications for training, education or human resource management – and which would have presented an efficiency/effectiveness-related explanatory factor – but instead on moral, responsibility and accountability factors, especially for the commander. A second difference between the original NCW concept and NEC concerned the style of the NEC sub-divisional paper document (German Ministry of Defence 2006a), which was comparatively intellectual and theoretical in character. It was argued above that there was a culturally grounded dislike for undertaking what would be considered to be a quick solution to a problem. Apart from this, the abstract conceptualisations of NEC were the result of a reluctance to either adapt the concept to the most likely operational challenges or the inability to link it to a national military strategy or doctrine, as this was constrained by the political level. As will be discussed below, the implementation of this intellectual and conceptual NEC concept caused problems, too.

This sub-section summarised the findings on the utilisation of the U.S. NCW concept. Process-tracing revealed that the Bundeswehr took considerably longer to draft a joint NEC concept – compared to other transformational concepts outlined in the 2004 Conception of the Bundeswehr and to similar processes in the U.S. and in the UK. Furthermore, and despite the different operational challenges for the U.S. and Germany, NEC largely resembled U.S. NCW. A number of factors – all representing instances of institutional legitimacy – have been identified that explain the slow pace of conceptualisation and the comparatively high degree of concept faithfulness.

5.1.3 Implementing NEC in Germany

A robust implementation scheme for NEC did not exist until 2010. With respect to implementation, the NEC sub-divisional paper was rather vague. Except for the common operational picture and references to stand-off precision weapons, joint fire support, and blue force tracking, the document did not refer to the specific capabilities to be achieved through NEC. A clear definition of what the initial operational capability (IOC) or even a full operational capability (FOC) comprised of was also missing and, thus, no connection was made between NEC conceptualisation and NEC implementation (German Ministry of Defence 2006a: 10). Finally, for years into the programme no time frame was announced as to when which state of NEC readiness should be achieved.

NEC governance structures were decentralised and hampered NCW utilisation. Furthermore, the Armed Forces Staff had proposed a roadmap for NEC; however, the services opposed the idea arguing that this would increase the pressure on a force already strained by the deployment in Afghanistan.[98] Without a comprehensive milestone plan or roadmap, NEC implementation was in the hands of the procurement branch. Comprehensive strategies for other NEC parts such as information management, training or human resource management were missing. Between 2004 and 2009 NEC implementation was untargeted. Only in 2009 the Chief of Federal Armed Forces Staff requested the demonstration of a NEC IOC in the form of an exercise (DemoEx) by 2013. This eventually resulted in the first armed forces-wide effort to implement NEC. However, the DemoEx did not receive unequivocal support. Critics complained the demonstration would bind resources urgently needed in the ISAF operation. The demonstration scenario was based on a solely joint and national scenario that did moreover not reflect German military reality in Afghanistan, which was marked by multinational cooperation. The following sub-sections on the implementation of NEC provide an overview of the NEC programmes, the governance structure and the actual application of NEC. Following this, the concluding section summarises the results.

5.1.3.1 NEC Programmes

In spring 2009 the Armaments Directorate in the Ministry and the Federal Office for Equipment and Procurement (BWB) issued an analysis of the network-readiness of the Bundeswehr's weapon systems. While Navy and Air Force systems met the requirements, Army systems mostly lacked the ability to be integrated into a joint C2-network. The report showed that, out of 70 weapons projects, only 6 were network-ready. 11 projects had some limited NEC capability. The rest was not network-ready at all. Moreover, radio communication across Army weapon command systems – for example between a helicopter and a protected vehicle – was often not possible.[99] Similarly, an

98 Interview, Bonn, 20 February 2009.

99 Presentation given by an official of the Armaments Department at the NEC Course at the German Armed Forces Command and Staff College, Hamburg, 10 June 2010. Also see Lange (2005).

Army report on procurement projects concluded that NEC initial operational capability was unlikely to be achieved before 2015 (Rösner 2006).

Despite these critical assessments, a number of NEC projects had been implemented. Compared to the introduction of NEC in the UK, the German implementation of NEC, however, is harder to assess because a clear implementation plan, and thus an overall network architecture, was missing. Nonetheless, a number of NEC-relevant equipment projects could be identified. These projects can generally be divided into three categories: communication systems, command and control systems and reconnaissance and surveillance systems.

In 2006 the contract for a military satellite communication system (SATCOMBw2) was signed and two communication satellites were launched in 2009 and 2010, thus reducing the Bundeswehr's reliance on commercial satellite capabilities. MobKommSysBw (signed in 2007, roll-out starting in 2009) is the mobile communication system that provided tactical data transfer. The trunked radio system Tetrapol was already in service with approximately 10,000 radios and was also planned to have the first software-defined radios rolled-out by 2012. With regards to communication systems, the NEC IOC was achieved in 2010, a staff member of the MoD Directorate for Modernisation noted.[100]

In the realm of C2, two major projects were initiated – one was civilian in nature, the other military. Reflecting the constitutional separation between the military and civilian administration, there had been two separate modernisation efforts both in the areas of *white* (civilian) and the *green* (military) IT. In the area of white IT and in an attempt to modernise the Bundeswehr's whole information infrastructure, the MoD committed in 2006 7 bn Euro to a 10-year programme called Herkules. Within the framework of a public-private partnership, the Bundeswehr, in cooperation with a consortium (Siemens and IBM), started setting-up a fiber network linking fixed military sites in Germany, integrating all Bundeswehr computers into one national system. Herkules thus provided the backbone of the Bundeswehr's information network. In the area of applications, the Bundeswehr implemented SAP standard software (SASPF), which replaced the myriad of proprietary software programmes (Ackerman 2009b).

The development of a *green* joint and interoperable Command and Information System (CIS) in the military realm, however, proved to be problematic. Due to financial restrictions, technical complexity and current operational requirements, existing command and information systems, notably the Army FüInfoSysH, the Air Force ACCS and the Navy MCCIS, had not been replaced with one joint system.[101] Instead, a joint system (FüInfoSysSK) was developed in parallel to the services' CIS upgrades. Only in a very basic version, the joint CIS was deployed to Afghanistan. It might take years though until the full version can link into the existing individual services and coalition CIS will be deployable (Wiegers 2010). This problem was multiplied for the Army, because the Army CIS struggled to demonstrate interoperability not only with the joint CIS but also with existing Army battle management systems. As a result, the Bundeswehr deployed several CIS to Afghanistan that lacked necessary interoperability.

An interesting development occurred in the summer of 2010. In 2008 the U.S. demanded that NATO allies in ISAF should contribute to the Afghan Mission Network

100 Presentation given at the NCO Course at the German Armed Forces Command and Staff College, Hamburg, 7 June 2010.

101 See Bundeswehrplan 2007 (German Ministry of Defence).

(AMN) – a multinational command, control, computer, communication, intelligence, surveillance and reconnaissance (C4ISR) network for all ISAF forces (Collins 2010). In January 2010, NATO requested Germany to present an interoperable CIS solution for the AMN by 10 July 2010. As the German CIS could not provide the interoperable system, the Ministry decided to lease an off-the-shelf CIS (FOC+) developed by Thales. This decision was spurred by the announcement of NATO HQ to deploy two U.S. brigades to the Regional Command North, which was under German command.

In the area of world-wide reconnaissance, the most important project was the satellite system SAR-Lupe that became operational in 2007. SAR-Lupe consisted of five small all-weather satellites and represented the state-of-the-art in radar technology. The project as such was a direct result of the Kosovo War (1999) experience when German requests for U.S. satellite imagery were not always positively answered. Area-wide reconnaissance systems included the EURO Hawk operated by the German Air Force as well as the so-called Recce Lite pod, which was fitted into manned Tornado Aircrafts, with a number of small tactical drones being deployed in Afghanistan and Kosovo as well. To provide for a more sustained tactical surveillance in the northern Afghan region, Germany opted for leasing the Heron 1 UAV from Israel in March 2010 as an intermediate solution for the theatre-depth imaging surveillance system SAATEG (Knittlmaier 2010).

There are some more general observations concerning defence spending and procurement. Significant amounts of the investment budget were bound in long-term projects that did not necessarily contribute to the current security challenges. In the year 2009, for example, only 7 percent (375 mio Euros) of the investment budget were free to be spent on new projects (Forster 2009). The roll-out of Cold-War systems like the Eurofighter and the attack helicopter (originally planned as an anti-tank helicopter) Tiger only started in the new millennium. These and other legacy systems still reflected an overcome territorial defence doctrine. Some of these systems were relabelled as NEC projects by the services (Klos 2006).

Some controversial procurement decisions were taken with a view towards supporting the German defence industry. The *Bundeswehrplan 2009*, for example, explicitly linked some procurement projects (Leopard2, Infanterist der Zukunft, Eurofighter and underwater vehicles) to their importance in supporting the German defence industry. A telling episode was the struggle to create a UAV capability in Afghanistan within the framework of the SAATEG programme: In 2002 the Bundeswehr identified a capability gap with regards to airborne ISR. Yet, due to the quite different views between the Ministry (which preferred a national solution) and the Air Force (which was in favour of systems quickly available on the international market), it took five years for the MoD to invite bidders. While different views continued to exist after the bidding process started, two available systems were identified (the U.S. Predator B and Israeli Heron TP). However, the Ministry decided in 2009 to take neither and instead favoured the EADS proposal to develop a UAV in the following years. This, of course, would have meant a further delay of several years, which led to an outcry in German security circles, as UAV were considered an important force protection tool. Not purchasing an available solution while waiting years for the roll-out of a national product was interpreted as a "deadly absurdity of German armaments policy".[102] In the end, due to the worsening security

102 http://www.geopowers.com/Machte/Deutschland/Rustung/Rustung_2009/rustung_2009.html#SAATEGV (accessed 4 April 2009).

situation in Afghanistan, the Ministry opted for an 'intermediate' solution later in 2009 by leasing three Heron drones from Israel for deployment in Afghanistan.

A further problem was bringing NEC to 'the last mile'; from the command posts and the units in the theatre. Despite an overall increase in investment in information technology, almost 80 percent of this investment was devoted to white IT, delaying necessary development of deployable CIS.[103] This situation was further complicated as IT procurement rested not with one but with two institutional branches in the ministry – the Directorate General for Armaments, which was responsible for armament projects, and the Modernisation Directorate, which was responsible for Herkules and all other IT-issues. Clearly, this two-tier structure of responsibility caused interoperability problems at the nexus of military and civilian information systems.

Finally, there were no joint concepts for information management and NEC training. An overall personnel policy with a view towards increasing the numbers of IT personnel or operators was also missing.

To conclude, implementation of NEC progressed well in the communication networks and reconnaissance and surveillance capability areas; yet an overall strategy for NEC was missing and the internal assessment of the Armaments Department of 2009 showed that there was neither an adaptation nor prioritisation of procurement projects, nor was there a general adaptation of the procurement process to address the challenges of NEC.[104] Thus, a number of implementation problems occurred: The Bundeswehr struggled to implement and deploy a coherent joint and coalition-interoperable CIS; the relation between civilian and military IT seemed unbalanced and not focused on urgent, operational demands; finally, armed forces-wide strategies for information management, human-resource management and training were missing. In addition, there were bureaucratic hurdles to an efficient NEC programme: Both the Directorate for Modernisation in the Ministry and the Directorate General for Armaments dealt with the acquisition of IT systems. This, quite naturally, resulted in the incoherency of NEC projects. The constitutional civil-military divide further fuelled this problem as it prevented the necessary transfer of knowledge between military staff having operational experience and civilians working in the acquisition branches. The next sub-section will show, moreover, that there was not only a structural inconsistency between the acquisition branches, but that the overall governance structure for NEC was decentralised and dispersed which further limited a resolute implementation of NEC.

5.1.3.2 NEC Governance

The NEC sub-divisional paper assigned the conceptual development of NEC to the department VI (planning) of the Armed Forces Staff (FüS VI). However, no new institutional section had been established within the FüSVI to deal with the new task of overseeing the NEC programme. Instead, NEC governance was a matter of coordination between different stakeholders within the Ministry. The ultimate responsibility for the Bundeswehr Transformation resided with the Deputy Chief of Staff. Transformation decisions were taken in the ministerial Transformation Coordination Group (TCG). In the TCG department chiefs from within the military staffs and the civilian departments met

103 Interviews, Bonn, 19 and 20 March 2009.

104 Presentation given by an official of the Armaments Department at the NEC Course at the German Armed Forces Command and Staff College, Hamburg, 10 June 2010.

every 3 months. TCG meetings were prepared by the Transformation Working Group (heads of division), which could delegate work on NEC issues to a NEC Working Group consisting of desk officers from the service staffs and civilian directorates.

The Transformation Centre and the Force Headquarters in Ulm were put on call for the Armed Forces Staff to support further conceptual development of NEC. The Transformation Centre contributed to the conceptual development of NEC by conducting or accompanying a number of military experiments. When in 2009 the Chief of Federal Armed Forces Staff requested the demonstration of a NEC IOC by 2013, the Force Headquarters was ordered to organise the actual conduct of the demonstration. The Command Staff within the Ministry was tasked with integrating the German CIS into the Afghan Mission Network.

This cursory examination of NEC governance structure showed that it was highly decentralised and dispersed; neither was there an authoritative section for the implementation of NEC in the Ministry, nor was there a military transformation command. In 2006, when the Transformation Centre was ceremonially inaugurated, then-Defence Minister Jung delivered a speech in which he declared the Centre was "the German equivalent to the U.S. Joint Forces Command, the American functional Transformation Command" (Jung 2006). Yet, the Centre did not receive adequate competencies, command authority or budget allocation responsibilities to live up to this ambitious statement. Four years later military officers therefore critically concluded that only the *idea* of a transformation was taken over from the U.S. but that actual implementation lacked dedication and ambition.[105]

5.1.3.3 NEC Application

A final point for assessing the implementation of NEC is the actual application in military operations, especially in Afghanistan. It is important to note that until 2008 the mission reality of the Bundeswehr in Regional Command North was different from the more violent South. Even though the security situation had worsened, the Bundeswehr still regarded its ISAF commitment as a *stability operation*. In 2008, a study on the Bundeswehr deployment in Afghanistan concluded that the Bundeswehr contingents lacked sufficient reconnaissance systems, did not have the means to achieve comprehensive situational awareness and were in need for an improved joint C2-capability (Lange 2008).

Officers nonetheless agreed that the equipment of the Bundeswehr in Afghanistan was "not too bad" and that, with regards to NEC, a lot of "muddling through" would take place with a view towards finding pragmatic solutions for the soldiers deployed.[106] Over the last years, a number of ISR solutions (Tornado Recce Pod and the Heron UAV) had been deployed to Afghanistan to provide real-time ISR. Yet the C2 problem remained.

---

105 Discussion during an NEC Course at the German Armed Forces Command and Staff College, Hamburg, 7 June 2010. Also see the press statement by the Commander of the Operations Command, Gen. Glatz on 10 April 2010 at the Henning-von-Treskow Base (http://www.bundesregierung.de/Content/DE/Mitschrift/Pressekonferenzen/2010/04/2010-04-10-potsdam.html, accessed 19 July 2010).

106 Yet, this improvisation on the ground was also criticised by the Parliamentary Commissioner for the Armed Forces (See: www.spiegel.de/politik/deutschland/0,1518,684005,00.html) (accessed 16 February 2011).

As a result of this slow NEC implementation, the Armed Forces Staff, and ultimately the Chief of Federal Armed Forces Staff Schneiderhan, in an effort to change the implementation process ordered in spring 2009 that the German NEC capability hat to be demonstrated in a joint demonstration exercise (DemoEx) in 2013 (German Ministry of Defence 2009). For two years, the Concept Development and Experimentation unit within the Transformation Centre, the single services and the Force Headquarters in Ulm were involved in preparing the exercise. Senior officers were quite critical of the actual usefulness of this demonstration, as it would have bound manpower resources and equipment urgently needed in Afghanistan.[107] In 2011, when it became apparent that most of the network equipment planned for the demonstration would not be ready for use, the DemoEx was cancelled. Tom Dyson moreover found that "doctrinal disagreement between the individual services over the implications of networking for command and control" and also "the need to integrate the lessons of participation in the Afghan Mission Network (ISAF's C4ISR network)" added to the decision to call off the DemoEx (Dyson forthcoming, 2013). What came as a relief probably to most units involved in preparing the exercise shed – again – some light on the actual achievements in the German NEC programme.

5.1.3.4 Conclusion

The previous sub-sections traced the NEC-implementation process in Germany. Some progress had been made in procuring ISR and C3 assets. Yet, as neither a comprehensive implementation roadmap nor a forceful governance structure existed, a number of important issues remained untouched that would have been decisive for the successful introduction of NEC: information management, training and human resource management. This concluding section will assess the variables *implementation pace* and *faithfulness of implementation* by which the overall adoption of NEC can be assessed and assumptions concerning efficiency/effectiveness and institutional legitimacy as drivers for NEC can be made.

Table 13a: The Implementation Pace of NEC in Germany

Pace of Implementation	Slow
Explanatory Factor	**Nature of the Factor**
Lack of Centralised Authority/Consensus Principle/ Joint Underwriting	Institutional Legitimacy
Failure to Formulate an Implementation Plan	Institutional Legitimacy

The pace of implementation was slow and the faithfulness of implementation – that is the relation between the officially expressed intent to introduce NEC and its actual implementation in terms of resources allocated, delivery of equipment and the changing of procedures – was low, too.

107 www.spiegel.de/politik/deutschland/0,1518,684005,00.html (accessed 16 February 2011).

Table 13b: The Implementation Faithfulness of NEC in Germany

Faithfulness of Implementation	Low

Explanatory Factor	Nature of the Factor
Decoupling of NEC from Operational Pressure	Institutional Legitimacy

Despite the enthusiasm for NEC at the military leadership level and the decision to make NEC the core feature of the Bundeswehr transformation, implementation was slow compared to the British case. After the release of the NEC sub-divisional paper, no milestone plan or roadmap was drafted and it took nearly three more years until the Chief of Federal Armed Forces Staff set a date for an IOC demonstration. Thus, commentators such as Elianor Sloan might have a point in asserting that most of the emphasis on German military transformation was rhetorical (Sloan 2008: 72).

Implementation, furthermore, appeared to be unguided to the point where it partly contradicted the original aims of the transformation process – improving jointness, multinational interoperability and mission readiness. Assessing the transformation progress in 2008, Chief of Federal Armed Forces Staff Gen. Schneiderhan pointed at the difficulty of achieving service jointness in operations. The problem, he stated, was not one of technology, but of culture (Schneiderhan 2008). With regards to multinational interoperability, the deployment of a joint CIS that is be able to link into NATO networks might still take several years. In the case of the Afghan Mission Network, Germany only reluctantly, and due to pressure of the U.S., turned towards an intermediate commercial solution. Similar problems appeared in the case of ISR assets (SAATEG, see page 122 above) that were urgently needed by troops fielded in Afghanistan.

Apart from the difficulties in achieving the more general transformation goals also specifics of the German NEC approach – notably the focus on the individual and the announcement that command processes would be revised (see page 113 above) – had not been put into practice either.

Why was NEC implementation so slow? As NEC had been the core concept of German military transformation, and as this transformation was envisaged to increase mission effectiveness, it is puzzling that the Bundeswehr did not manage to put its ambition into practice.

One set of factors that considerably slowed down the implementation process was the lack of a centralised authority during the implementation process. The Chief of Federal Armed Forces Staff did not use his powers that had been extended in 2005 to forcefully push NEC implementation. The Armed Service Staff did not manage to enforce the NEC vision either. Instead, the service staffs and the civilian ministerial directorates negotiated transformational matters on the basis of the principle of consensus. This was – against the background of the absence of a joint understanding of NEC and different stakeholder interests – almost a guarantee for deadlock and decisions resembling the lowest common denominator. As a result, the services did not agree upon a roadmap or a comprehensive implementation scheme. Consequently, a number of programmes in the procurement area simply continued to run without being integrated into an overall strategy. Additionally, no new programmes in the areas of information management and NEC training were set-up. The Armed Forces Staff was singled out by the 2004 Conception of the Bundeswehr to lead the transformation. In order to maintain institutional legitimacy, however, neither the Armed Forces Staff nor the strengthened

Chief of Federal Armed Forces Staff changed internal decision-making processes or established new government structures for NEC.

The low level of faithfulness of implementation is best explained by the *decoupling* between NEC and operational reality. Under this umbrella, a number of related factors can be summarised: The failure to refocus budget priorities in favour of deployable CIS and the failure to mobilise the Bundeswehr to support NEC by conveying the usefulness of the concept for current military operations. It seems that budget priorities were set in favour of national solutions (SAATEG) or – in the case of Herkules – with a focus on the civilian part of the Bundeswehr. In both cases, the reluctance of the military leadership to change this course is an indicator for its desire to maintain institutional legitimacy vis-à-vis the institutional environment (here, national defence industry and political leadership) instead of insisting to increasing military effectiveness.

NEC did not receive much support from the services. The Armed Forces Staff did not tackle this issue by *explaining* the use of NEC for current deployments. Strangely, even during the implementation phase, the theoretical orientation of NEC remained strong, which culminated in the decision to demonstrate NEC using a scenario that did not resemble mission reality. As this demonstration was furthermore perceived as a burden for the participating contingents, implementation did not only fail to bring about mission orientation. By binding scarce resources, it ran counter to its original aim. The decoupling between aspiration and reality was an expression of schizophrenia of German defence that is still uncomfortable engaging in combat operations.

The previous subsection showed that NEC utilisation was slow and detached from the original, or officially expressed, aim. The hesitation to come to terms with NEC continued during the implementation phase, which a senior officer summarised in 2010 by stating: "When it comes to NEC, the Bundeswehr has gotten completely stuck".[108]

5.2 NEC in Germany: Concluding Remarks

For a long time, information on the German variant of Network-Centric Warfare was rare. As mentioned above, the Bundeswehr, especially the Air Force Development Centre, offered some pamphlets about the concept on their website. In 2007 Sabine Collmer had published a paper entitled *The Influence of RMA and Net-Centric Operations on the Transformation of the German Armed Forces* (Collmer 2007). In 2010, Heiko Borchert contributed a chapter, concerning NEC, to an edited volume that sought to compare the various military transformation projects in selected European NATO member states (Borchert 2010). The absence of a communication strategy concerning NEC on the part of the Bundeswehr and the relative neglect of the concept by academics and security experts is, at first sight, surprising, as NEC was officially declared the core feature of the German military transformation in 2004. Yet, apparently, the introduction of NEC was (and still is), to some extent, at odds with the aims of German defence policy and operational realities and, thus, did not raise the awareness the officially expressed determination to implement the concept would have suggested.

108 Discussion during the NEC Course at the German Armed Forces Command and Staff College, Hamburg, 7 June 2010.

Table 14: NEC Adoption Patterns in Germany

Timing	Late
Adoption Pace	Slow
Concept Faithfulness	High
Faithfulness of Implementation	Low

Starting with the more descriptive task of uncovering the unique adoption patterns that underlay the adoption process this chapter illustrated that Germany adopted NEC late in 2004 and took an exceptionally long time to utilise the concept. Implementation was unfocused, lacked guidance and failed to materialise NEC. In a second step the aim of this research was to find explanations for the unique adoption patterns, especially in comparison with the British case of NCW adoption. What, then, were the driving forces that accounted for the late and slow development of the NEC Programme that even among those officers who were working within the NEC domain was considered being rather unsuccessful?

Qualitative research revealed that, throughout the adoption process, legitimacy-related factors accounted for the pace and substantive outcome of the NEC programme. Critical Triggers for NCW adoption in Germany are summarized in the following table.

Table 15: Drivers of NEC Adoption in Germany

Level	Efficiency/Effectiveness	Institutional Legitimacy
International	Operational Requirements (Incentives)	US Request for Transformation (Incentive) NATO Transformation and NATO NEC (Incentive)
Domestic		Lack of General Interest/Support for Armed Forces' Combat Role (Constraint) Military Transformation Project (Incentive) Unclear Political Guidance for Armed Forces' Role (Constraint)
Organisational		Interservice Boundaries (Lack of Jointness) (Constraint) Mission Command Discussion (Constraint) Conceptual Approach towards Innovations (Constraint)

It has been discussed above that adoption processes are influenced by factors that can be arranged along two dimensions: first, the respective environmental level (international, domestic, organisational) and second, whether the factors related more to rational or sociological incentives or constraints. The story of NCW adoption in Germany illustrated three observations. First, the introduction of NEC in Germany was triggered by and conducted under the influence of legitimacy-related factors. Rhetoric aside, no evidence could be found that a thorough assessment of the current or future operational Bundeswehr needs had been at the core of NEC. This is not to say that single equipment projects especially in the area of communications lacked this perspective, yet an overall strategy for the NEC Programme was missing between 2004 and 2010. Second, in the case of Germany the domestic level provided the most critical factors that steered the NEC adoption process. A third observation is the presence not only of incentives but also of factors that were constraining in nature, and that hampered the implementation of the concept.

After years of neglect NCW received attention in Germany in 2003. It had been chosen to lead German military transformation (mostly a political project), when the international *hype* over NCW was high. When the new Minister of Defence Struck struggled to justify new reform steps *transformation* and *NEC* appeared as a ready-made conceptual package that was likely to gather fresh internal support for reform. Moderate pressure to adopt NCW was exerted through NATO (first through the NATO transformation, later trough the ISAF mission). Although these capability improvements would have represented an increase in German military effectiveness, the initial decision to adopt NCW appeared rather as an interplay between international expectations and a domestic window of opportunity. Both reasons represent instances of increasing institutional legitimacy, first vis-à-vis NATO and the U.S., and then in relation to the German public and the political level.

The failure to bring NCW to life despite the announcement that NCW was the pillar of German military transformation was the result of a strategy of decoupling of policy formulation and decision from the actual implementation.[109] Decoupling is a concept found in the literature of organisational institutionalism and initially described a strategy of organisations to separate "their formal structure from their production activities when institutional and task environment are in conflict" (Boxenbaum/Jonsson 2008: 79). Decoupling can occur in two ways. First, it might be a response by an organisation if external demands contradict internal structure (Meyer/Rowan 1991). Second, decoupling might be chosen if the environment is confronting the organisation with multiple, contradicting demands. In the case of German NCW adoption the latter occurred. Apparently, in order to reassure both NATO, which demanded a more active German military role and modernised equipment, and the German public, which objected a combat role for the Bundeswehr, Germany had chosen a strategy of decoupling NCW conceptualisation from implementation.

In sum, the NEC project was not adapted to best suit the deployment of German soldiers abroad, especially in Afghanistan. The Bundeswehr's strive for maintaining domestic institutional legitimacy instead of increasing military effectiveness and efficiency is the most plausible way to explain the late timing, the slow pace and the low degree of faithfulness of implementation of NEC. Therefore and in stark contrast to the UK, in the case of NEC adoption, Germany was a legitimacy maximiser.

109 By referring to a *strategy* of decoupling an agentic perspective is taken on organisations that are supposedly capable of responding differently to external demands (Oliver 1991).

6 Conclusions and Implications

This book examined the adoption of Network-Centric Warfare (NCW) in the British and German armed forces between 2001 and 2010. NCW was developed by the U.S. military during the late 1990s and, in 2001, became the cornerstone of a profound military transformation that was pursued by the incoming U.S. Secretary of Defense, Donald Rumsfeld. Since 2001, a number of states followed the U.S. example and introduced NCW in their armed forces, albeit with variations in the conception of NCW, ambition and, eventually, success. Empirical research revealed considerable differences in the adoption processes between Britain and Germany. The main argument of this study is that these variations in NCW adoption can partly be explained by the existing differences in the respective institutional environments of either the British or the German military organisation (and their responses to these environmental factors). Accordingly, during the empirical research, attention was paid to the presence or absence of predefined environmental factors located either at the international, societal, domestic political or organisational level and how the military organisations reacted to these factors. This concluding chapter will review the empirical findings against the theoretical framework provided. Furthermore, the wider implications that this research has for both the academic fields of military change and diffusion research, and the practical relevance of the findings for armed forces in transition, will be discussed.

6.1 NCW Adoption Patterns Compared

Having traced the introduction of NCW in the UK and Germany it appears that in comparison both states differed considerable in the temporal dimension (timing/pace), concept faithfulness and the faithfulness of NEC implementation. NCW adoption in the UK started early in 2002 and proceeded comparatively quickly. Together with Sweden and Australia, which had been in the process of introducing NCW since 2001 and 2003, respectively, the UK belongs to the group of *early adopters* of NCW. The German Ministry of Defence, in contrast, only declared in 2004 that it would introduce NCW. Thus, depending on the reference framework, Germany can either be considered to be average with respect to timing or even be labelled a *late adopter* of NCW. The difference in timing between the UK and Germany grows even wider when the *pace* of adoption is considered. The UK introduced NCW not only earlier, but also more quickly than Germany.

Table 16: NCW Adoption Patterns in the UK and Germany (2001–2010)

	UK	Germany
Timing/Pace	Early/Quick	Late/Slow
Concept Faithfulness	Low (Selected)	High (Copied and Modified)
Faithfulness of Implementation	Moderate (–2007) to High (2007–)	Low

Further differences between the two cases exist with regards to the adoption outcomes. In terms of variation from the original NCW concept (*concept faithfulness*), empirical research revealed that the UK picked certain features from the U.S. concept and adapted the British Network-Enabled Capabilities concept to its needs, whereas the German *Vernetzte Operationsführung* largely represented a copy of the original NCW concept.

This resulted in a number of discrepancies between the German NEC concept and the general lines of German defence policy. The German NEC concept, for example, was lacking a multinational perspective. Yet, due to constitutional limitations, government policy laid down in the Defence White Book (2006) and practice, Germany deploy forces abroad – with the exception of rescue missions – only in multinational settings. The conceptual emphasis on the whole spectrum of military action (reconnaissance, command and control and engagement) is also astonishing when considering the fact that the Bundeswehr mission reality is rather on stabilisation operations and not on combat missions.

The third analytical adoption category is *faithfulness of implementation*. Here, it was shown that despite the setbacks in implementation until 2007 the UK not only implemented NEC quickly, but that implementation then followed the conceptual underpinnings of NEC. The re-adjustment of the programme in 2007 to emphasise training and information management, and the re-adjusted governance structures reflecting the conceptual building blocks of NEC, are the most visible manifestations of the close relation between conceptualisation and implementation in the UK. In the case of Germany, in contrast, implementation was at odds not only with the concept, but also with the aims of German defence policy and operational realities.

6.2 Substantial Differences

In comparison a number of substantial differences between the NEC conceptualisation and implementation became apparent. These are summarized in the table below.

Table 17: British and German NEC Compared

	UK	Germany
NEC Concept Focus	Coalition Interoperability, Jointness, Interagency Interoperability	National Jointness
NEC Implementation Focus	Deployed Forces	Demonstration Exercise 2013
NEC Implementation Areas	Equipment, Training, Information Management and Doctrine	Equipment
NEC Governance	Integrated, Centralised and Authoritative Structures with Monitoring Capabilities Mirroring NEC Concept	Decentralised Structures not Relating to NetOpFü Concept
Verification	Combat Evaluation	Demonstration Exercise 2013
Programme Relevance	Increasing Military Effectiveness	Military Transformation Project
Knowledge Acquisition	MoD Structures	Entrepreneurship
Role of U.S. in Adoption Process	Information Sharing	Persuasion

The aim of this study, however, was to go beyond the description of the national adoption processes and outcomes. Rather, it sought to establish an explanatory model that could account for the observed differences. Thus, the next section discusses the explanatory model in the light of the empirical results.

6.3 Discussion of Research Questions and Argument

Using an institutionalist framework, the research argument was that military organisations function within an environment that sets constraints or gives incentives to their actions. Thus, the NCW adoption process and its outcomes would be influenced by demands from the environment (and how the military organisation reacts to these demands). In order to operationalise this understanding of exogenous institutionalist factors to concept adoption, a set of potential factors was developed spanning a number of environmental levels relevant for military organisations – the international level, the societal level and the domestic political level. Although the focus of the empirical research was on the exogenous factors, the organisational level (being the basis for military culture) was also considered an important aspect and was thus included in the framework.

On each of the aforementioned environmental levels, it was then assessed, whether the military organisation was more responsive to demands that were either related to questions of military effectiveness and efficiency or institutional legitimacy. Unarguably, military organisations are not perfect instruments of rational choice, but differences exist between the British and the German military as to what extent the two organisations rather followed a functionalist imperative in contrast to being subject to other external pressures.

Research revealed that, in the case of the UK, the NEC concept was mainly introduced with a view towards increasing the British Armed Forces' military *effectiveness*, while in the case of the German Bundeswehr, NEC was rather adopted out of *legitimacy* concerns. Hence, in the latter case, concept adoption was guided by the aim of maintaining or increasing the Bundeswehr's institutional legitimacy.

As the table below shows, in the case of the UK there were mainly incentives from the international levels at play that could be grouped in the effectiveness/efficiency column. In contrast, empirical research in Germany uncovered the presences not only of incentives but also of constraints to NEC that stemmed both from the international and domestic level and could be attributed to increasing or maintaining institutional legitimacy.

In the case of the UK, NEC adoption was mainly caused and influenced by the need to stay, or become, interoperable with U.S. and coalition armed forces in the theatre. Furthermore, NEC was seen as a way to save resources and, thus, to increase the British Armed Forces' efficiency. Although interoperability with U.S. forces was the main concern, joint interoperability was also still an issue. All in all, the case of NEC adoption in the UK was mainly shaped by an emphasis for increasing military effectiveness or organisational efficiency. As the British Defence White Paper stated, "NEC is crucial to the rapid delivery of military effect" (British Ministry of Defence 2003a).

Table 18: The Drivers of NEC Adoption in the UK and Germany

	UK		Germany	
Level	Effectiveness/ Efficiency	Legitimacy	Effectiveness/ Efficiency	Legitimacy
International	Need for Interoperability with U.S (Incentive) Need for Coalition Interoperability (Incentive) Operational Requirements (Incentive)	Ambition/ Prestige (Incentive)	Operational Requirements (Incentive)	US Request for Transformation (Incentive) NATO Transformation and NATO NEC (Incentive)
National (Societal)		General Interest or Support for Armed Forces' Combat Role (Incentive)		Lack of General Interest/Support for Armed Forces' Combat Role (Constraint)
National (Political)	NEC as a Means to Save Resources (Incentive)			Military Transformation Project (Incentive) Unclear Political Guidance for Armed Forces' Role (Constraint)
Organisational	Need for Joint Interoperability (Incentive)	Pragmatic Approach towards Innovations (Incentive)		Interservice Boundaries (Lack of Jointness) (Constraint) Mission Command Discussion (Constraint) Conceptual Approach towards Innovations (Constraint)

Evidence seems to confirm the assumptions found in the literature that early adopters are driven rather by considerations of efficiency while late adopters introduce a foreign concept to increase their institutional legitimacy (Eyre/Suchman 1996: 95; Goldman/ Ross 2003: 375). Early adopters make an 'optional innovation-decision'. Late adopters, in contrast, have 'followed the herd' making a collective innovation-decision (Rogers 2003: ch 1).

The German introduction of NEC was motivated, and shaped, by factors that could be subsumed under the label 'institutional legitimacy'. Adoption was greatly influenced by contradicting institutional pressures from the organisation's environment. Since 2002, Germany has been committed to the NATO transformation process, which also pursued an own NEC initiative. Moreover, evidence shows, that there has been a moderate push from the U.S. administration to transformation. In 2003 and 2004, when the need to further reform the German Armed Forces increased, a window of opportunity opened for NEC. NEC was perceived as an innovative *en vogue* concept that could gather internal and external support for a renewed military reform attempt. Yet, as the motivation for introducing NCW did not originate from an internal demand for more effectiveness, the concept was later not adapted in a way that reflected the Bundeswehr's functional purposes, let alone its mission reality. Instead, NCW introduction in Germany was *decoupled* from the Bundeswehr's missions, that is its officially stated mission was not entirely reflected in the actual operational conduct. A telling example was the planned NEC demonstration exercise (DemoEx), which did not reflect multinational operational reality in Afghanistan as it was based on a national scenario.

As a result, it is argued that, in the United Kingdom, NCW was adopted to increase the military effectiveness and efficiency of the armed forces, whereas the adoption in Germany was mainly driven by the aim to maintain the armed forces' institutional legitimacy. Thus, in the case of NCW adoption, the British military was an efficiency maximiser whereas the Bundeswehr was a legitimacy maximiser.

6.4 Legitimacy of Efficiency and the Efficiency of Legitimacy

The finding that in the case of NCW adoption the UK was an effectiveness and efficiency maximiser, whilst Germany was a legitimacy maximiser, might raise the question of the relationship between legitimacy and effectiveness. Is what we understand as efficiency and effectiveness not in itself socially constructed (Powell 1991) and is the strive for efficiency not something that is also legitimate? Or, to turn the argument around, would it not be efficient to follow certain demands from the environment that, at first glance, do not seem effective, but in the long-term secure support from the environment? These are valid questions and, in complex societies, effectiveness, efficiency, and legitimacy undoubtedly overlap.

Both military organisations, however, serve certain defined purposes (crisis management and defence) employing a preset group of actions (reconnaissance missions, fighting, command and control, protection and support functions). While the mission statement is political and shaped by a number of factors that are not necessarily oriented towards effectiveness, the carrying-out of the mission should be focussed more on effectiveness and efficiency. It appears, however, that, in the case of Germany, this functional realm was also steeped in a struggle for legitimacy.

6.5 Civil-Military Relations and Strategic and Military Culture

As stated above, the purpose of this study was to offer a complement, not an alternative, explanation for the differences in the outcomes in cases of military concept adoption. Important overlaps exist with (and between) explanatory frameworks that focus on the role of strategic changes, military culture and civil-military relations. Changes in the strategic environment, i.e. the emergence of new threats, may result in military change. While the strategic environment can thus be an enabling factor for change, a nation's *strategic culture* might influence the way a military threat is assessed and responded to. German strategic culture, for example, only moves away from its former orientation toward territorial defence slowly and, despite the existence of threats to German security by terrorist networks, the 'culture of restraint' (Bredow 2007; Harnisch 2005; Hilpert 2010) is still prevalent. The British Armed Forces, in contrast, have been involved in out-of-area operation throughout the Cold War and still today reinforce this conception of having an expeditionary role (Vennesson et al. 2009).

Military culture is often seen as an impediment to military change and innovation (Farrell/Terriff 2010: 8). This seems to be true in the case of Germany, where a lack of joint understanding of the theoretical approach to concept development and risk avoidance hampered the adaptation of NEC. In the case of NEC adoption in the UK, however, the pragmatic, and "antiintellectual" fighting culture of the British Armed Forces (Nagl 2005 [2002]: 36f.) fostered innovation and change.

Finally, there are differences in the *civil-military relations* between the two cases, which might also have contributed to the differences in concept adoption. While in Germany the political level is generally cautious to empower the military elite with competencies to regulate military affairs, the British military is less subject to political influences regarding their internal structuring (Born/Beutler 2007). To use Huntington's terminology, the Bundeswehr, in comparison, has been subject to an ever-maximising civilian power (subjective control of the military) whereas the British Armed Forces could maximise their military professionalism (objective control of the military) (Huntington 1957: 83). As a result, the introduction of NCW in Germany was marked by strong civilian involvement (illustrated, for example, by the framing of NEC as a military transformation project) while, in the UK, NEC remained below public awareness and was considered nothing more than a military force enabler.

In this study's theoretical framework the strategic environment, strategic culture, civil-military relations and military culture were integrated into the explanatory model. Except for military culture, all other factors could be treated as exogenous to the military organisation, imposing institutional incentives or constraints to military change.

6.6 Decoupling

An inconsistency existed in the case of German NEC adoption: On the one hand, NEC conceptualisation stressed the concept's value for combat action. On the other hand the actual implementation of NEC resembled Germany's culture of restraint. This puzzling twist in the adoption of NEC can best be understood by referring to the concept of decoupling.

The Bundeswehr was confronted with the situation of two contradicting environmental demands: First, in the frame of NATO and against the backdrop of the war in Afghanistan the U.S. demanded in 2002 that the European allies would take their share of the military burden. For Germany this meant a changed equipment focus to become interoperable, being able to fight in a military coalition including the political decision to send combat troops abroad. Second, at the domestic level German defence was interested in the upkeep of the image of the non-fighting Bundeswehr in order to maintain domestic support for low risk crisis management operations abroad. Between 2001 and 2010 these two environmental demands could not be reconciled. As a result, the Bundeswehr chose the strategy of decoupling in order to increase its legitimacy at the international level and maintain its domestic legitimacy at the same time. The military did so by symbolising compliance with the NATO transformation agenda while maintaining a rather low combat image at home. As a result, formal policy announcement to introduce NCW was not followed through resulting in the finding of a low faithfulness of NEC implementation.

6.7 Implications for Further Research

While the present book has focussed mainly on establishing and explaining NCW adoption differences in the UK and Germany, other related aspects also merit closer attention.

6.7.1 Further Conceptual Work

As the previous subsection argued, the institutionalist perspective does not offer an alternative explanation of concept adoption and military change in general, but a complementary one. Yet, a robust research framework that specifies the relations between the institutional environment, strategic and military culture and civil-military relations is still missing. In today's academic debate about military change, indeterminacy seems to exist that does not allow to focus on all potentially important determinants of military change. However, taking into account the complexity of military organisations and their manifold relations, such a framework would, perhaps, not bring about generalisable inferences. Nevertheless, the areas of overlap between the existing approaches to military change need to be specified.

Especially within the neo-institutionalist school of thought, a number of concepts could be helpful in further understanding the dynamics of military change. Beyond the military culture argument, special attention could be paid to the concepts of path-dependency and critical junctures. For example, in the British case NEC adoption was influenced by the existence of earlier digitisation initiatives, the earlier introduction of the joint principle and the existence of central MoD digitisation programme structures.

6.7.2 Addressing the Role of the Innovator in Military Diffusion Processes

With a view towards further developing a framework of military diffusion, the relation between the innovator (in the case of NCW, the U.S.) and the adopters (here, the UK and Germany) – beyond the established unidirectional relation – could be worthwhile to look at. For example, did the British way of conceptualising and implementing NEC in turn influence the further implementation of NCW and its re-conceptualisation towards less techno-dependency in the U.S.?

6.7.3 Further Assessing the Spread of NCW

Another fruitful avenue for further research concerns the testing of hypotheses in a more quantitative setting. Apart from testing the validity of the institutionalist framework to address military change, such a quantitative view could also bring about insights about the global spread of NCW. As of 2010 a number of states have officially declared to introduce NCW: some of them with more, others with less success; some quick, others slow; some faithful to the original, others actively modifying the concept so to fit their own strategic or operational necessities.

Sweden, for example, had made its concept of Network-Based Defence (NBD) the foundation for the Swedish Armed Forces' modernisation process (Swedish Ministry of Defence 2004). NBD not only stressed interoperability with NATO (Ackerman 2007; Baddeley 2006; Kenyon 2007b). It also reflected the comprehensive understanding of Swedish security by enabling exchange of information and collaboration not only between military entities but also with other, civilian authorities involved in Sweden's total defence and crisis preparedness.

NCW in Australia is another example for concept modification as the concept clearly prioritised the Navy (Waters 2008: 20). This is plausible considering the strategic location of the 'island continent' in the Indian and Pacific oceans. Whereas Sweden and Australia modified the NCW concept, Denmark is a case where only certain features of the original concept have been selected. From the beginning, the Danish Network-Based Operations project (NBO) was rather mission-oriented, based on a set of military scenarios and closely connected to NATO's Network Enabled Capabilities initiative. Denmark did not emphasise the development of a joint national C2 system but instead favoured the use of NATO systems and standards such as M/CCIS (Maritime/Command and Control Information System).[110]

Table 19: Examples of NCW Variants (as of 2010)

	Concept Name	Start of the Programme	Main Focus of the Programme
United States	Network-Centric Warfare (NCW) Network-Centric Operations (NCO)	2001	Joint Interoperability for Expeditionary Warfare and Stabilisation Operations
Australia	Network-Centric Warfare (NCW)	2002/3	Networked Navy
Denmark	Network-Based Operations (NBO)	2004	Interoperability with NATO allies
Germany	Vernetzte Operationsführung (NetOpFü)	2004	Joint Interoperability
Sweden	Network-Based Defence (NBD)	2001	Joint Interoperability and Interoperability with NATO
UK	Network-Enabled Capability (NEC)	2002	Interoperability with the U.S. and Joint Interoperability for Expeditionary Warfare and Stabilisation Operations
EU	EU Network-Enabled Capability (EU NEC)	Draft 2008 and Ongoing Implementation Study (2010)	Comprehensive Approach
NATO	NATO Network-Enabled Capability (NNEC)	2003/2005	C2-Interoperability in NATO Missions (esp. ISAF)

Apart from national military organisations also NATO and lately the EU have developed some visions for network-enabled capabilities. It could be worthwhile to further assess the relation between success or failure of the respective national NCW programmes in terms of pace and implementation and the respective variances in the composition of efficiency/effectiveness and legitimacy drivers in each case. This would allow for more elaborate inferences about the impact of institutional forces at play during military diffusion and concept adoption.

110 See http://www.forsvaret.dk/FKO/eng/Defence%20Agreement/Pages/default.aspx (retrieved 5 May 2010).

6.7.4 Exploring the Strategic Consequences of Military Diffusion

Finally, and going beyond the effectiveness/legitimacy divide, the results of this study might raise the question of which strategic implications the diffusion of military innovations might have. Not only the UK and Germany, but in fact a large number of NATO members and non-members as well as NATO itself embraced the U.S. style military transformation by adopting concepts such as NCW and effects-based operations. With a view to the national strategic cultures does that mean that these states also adopted the 'American Way of War'? Did they perhaps even adjust their outlook on foreign policy and international security issues accordingly? Michael C. Horowitz addressed this question from a systemic angle in a recent publication (Horowitz 2010a). Focussing on the effects of military diffusion on the international system he finds that domestic adoption capacity (operationalised as financial and organisational capacity of a state to adopt a new military concept) accounts for the success or failure of a state to capitalise on a new military technology and, thus, for changes in the balance of power between great powers.

Against this background the question emerges of how the diffusion of military technologies or concepts influence strategic thinking, for example, whether or not the range of military capabilities at hand (here, NCW) determines the kind of crises scenarios a state is able to deal with. How seriously influence "existing organized capabilities (…) government choice" (Allison/Zelikow 1999: 176)? Could foreign concept-adoption lead to changes in national strategy through the back door? Another question concerns the expectations by others. By (symbolically) introducing foreign military concepts, even if they do not fit the national security strategy, a state might raise expectations by coalition partners, which might later result in unintended consequences. The current debate about Pooling and Sharing of military capabilitites in Europe might be an interesting case for assessing this question.

6.8 Policy Implications

Apart from the implications for further research, this study brought about a number of more practical inferences. Perhaps the clearest conclusion is that foreign concept adoption is not a solution to fix an existing lack of strategic debate. In the case of Germany, the transformation project and NEC came into being with no rigorous review of the purpose of the Armed Forces. Furthermore, NCW was not adapted to fit the existing operational reality or the broad strategic guidance provided by the 2006 Defence White Paper. This mismatch between NEC and mission reality resulted in a far-from-smooth implementation, a lack of internal support for the concept and wasted resources.

This kind of uncritical adoption might have severe consequences, especially for the armed forces operating under severe circumstances. In a recent study, Kjell Inge Bjerga and Torunn Laugen Haaland found that doctrine evolution during the 1990s in the Norwegian Armed Forces did not reflect operational experiences or strategic necessities, too. Instead, doctrines were detached from operational experience and rather determined by concepts evolving from the U.S. RMA debate and NATO doctrine (Bjerga/Haaland 2010). This neglect of national characteristics – be it military culture or national strategy – will most likely result in a drop of military effectiveness and also internal support for military change.

Thus, however intriguing a new military innovation might be at first glance, any foreign concept or technology with such wide-ranging effects on internal force structuring and procurement as NCW should be thoroughly assessed before political announcements to introduce them are made. Politicians and military senior officers should be careful before accepting new military concepts without any further consideration of national fit. Any introduction of foreign concepts should to some extent reflect the functional imperatives and organisational characteristics of the importing military. In the case of military adoption, *cherry picking* might be favourable to *faithful copying*.

A second conclusion is that operational concepts are not well suited to legitimate military reform projects. Linking innovations to reform projects might be a calculated decision, as innovative concepts might receive the attention of the professional audience, result in support of the reform project while at the same time working as an alibi to avoid difficult debates. Yet, in the end, the close linking of NCW and the military transformation project has at least twice – in the U.S. and Germany – led to a disregard of the network idea when the transformation programme came under attack. In Germany, NEC was equated with transformation, but transformation also meant base closures, relocating soldiers and, all too often, the network idea was considered to be yet another idea of the leadership to make life of the ordinary soldier harder.

A third conclusion is that all-out theoretical orientation is not worth aiming for when it comes to drafting military operational concepts or doctrines. Conventional wisdom suggests that it is not desirable to tinker long with concepts that have a practical relevance in military settings, which themselves are likely to change along a broad spectrum of scenarios. In order to bring concepts to life quickly, a pragmatic approach of constant adaptation might be better suited instead – an approach that favours many small steps of development instead of big jumps that might lead to results that are then harder to re-adjust. The announcement of the German Armed Forces Staff in 2009, for example, to have the NEC concept readjusted by 2013 – four years from the announcement – arouses the suspicion that the outcome of this revision will be yet another heavy-headed concept, that by the mere history of its time-consuming drafting will be even harder to change at a later stage. As the challenges of military conflict might change rapidly, military organisations have to be adaptable (admittedly, within the realms of strategic culture). Thus, the culture of doctrinal development should reflect this adaptability.

The British case of NEC concept-adoption was characterised by a number of factors that likely contributed to the success of the adoption: First, the new concept was linked to operational realities and thus quickly gathered defence-wide support. Second, there was a coherent logical connection between the assessment of operational needs, the specification of the NEC concept and its implementation, especially after the readjustment of the programme in 2007.

6.9 The Way Ahead

During the time of the writing of this book, a number of changes occurred at the political and military fields in the UK and Germany that are likely to affect the future course of the NCW programmes in both countries. In October 2010, the UK Government issued the Strategic Defence and Security Review (SDSR), which received high amounts of public attention as drastic cuts in terms of personnel and equipment for the British Armed Forces were announced (Her Majesty's Government 2010). Similar to the previous defence reviews, the document also elaborated on the missions and tasks of the British Armed Forces and, from this, derived the structure and capabilities of the forces. NEC, which was rather prominent in the 2002 and 2003 reviews, was not mentioned in the 2010 document. Instead, a new emphasis was put on cyber security. In 2011 the NEC vision was re-examined "to define what it aspires to achieve, and most importantly, what it can afford in terms of technology, information and its people" (Standen 2010: 81). The following quote from the recent MoD Information Strategy entails a brief assessment of NEC as well as the way ahead (British Ministry of Defence 2011): "MOD's approach to Information Superiority is also maturing. When Network Enabled Capability (NEC) was first conceived, its implementation required a revolutionary approach. Since that time, Through Life Capability Management (TLCM) and NEC governance structures have matured to the point where the concept could now be delivered as evolutionary change by Defence. As a consequence, Defence intends to update and simplify NEC (including its Handbook, JSP 777) and embed it as 'daily business' within Defence, alongside SDSR and Defence Reform Review changes".

It appears, furthermore, that compared to the previous reviews, less emphasis was put in the SDSR on the special relationship with the U.S. and interoperability issues. Supposedly, this was a political move to ease the factions in British society that are critical of the costly British engagement in Afghanistan (and earlier Iraq) alongside the U.S. Evidence suggests that after Tony Blair left office in 2007, some distancing from the U.S. has taken place. While some authors argue this was mainly "ritual distancing" on the diplomatic level while the strong relationship on the working level was probably not affected (Dumbrell 2009: 69), others see a diminished U.S. interest in the British partner (Kupchan 2010), make out a tendency of the British government to re-adjust its own foreign policy objectives, to seek more cooperation with European partners (Koydl 2010) and to focus more on domestic economics than on defence issues (Stephens 2010). Against this uncertain future of the Anglo-American defence relationship, it will be interesting to see if, in the near future, the focus of the NEC project will change away from the prime motivation of assuring interoperability with the U.S.

Similarly to the British armed forces, the Bundeswehr was faced with severe budget cuts totalling 8 bn Euros between 2011 and 2015. To what extent the NEC programme will be affected is not clear. In the current course of reform, now labelled the Reorientation of the Bundeswehr the *Transformation Centre* was turned into the *Bundeswehr Planning Office* and received more competencies especially in the realm of procurement. In a way, the NEC program was also upvalued. Since 2012 NEC was not anymore part of the Concept Development and Experimentation unit but was overseen from within an own unit, the "NetOpFü Dezernat" in the Planning Office. Yet, the future will show whether or not these institutional changes will give a new impetus to the German NEC.

Although the future of the NEC programs is slightly uncertain, the overall idea of gaining decisive military benefits through the networking of otherwise stove-piped information channels is not at all likely to disappear. In the UK as well as in Germany, the importance of smooth integration of C4ISTAR assets is widely acknowledged (see, for example, Codner 2010: 10), the protection of computer networks (cyber security) is currently moving into the focus of attention. Thus, all in all, the examination of NCW concept adoption in the UK and Germany provided both insights on the dynamics of military diffusion and concept adoption, and also on the highly complex technological and organisational settings military forces have to deal with in the information age.

7 List of Acronyms

4th Generation Warfare (4GW)
Afghan Mission Network (AMN)
Airborne Warning and Control System (AWACS)
Alliance between the UK, USA, Canada, Australia and New Zealand (UKUSA)
Allied Command Operations (ACO)
Allied Command Transformation (ACT)
Allied Rapid Reaction Corps (ARRC)
American, British, Canadian, Australian and New Zealand Army's Programme (ABCA)
Armed Forces Communications and Electronics Association (AFCEA)
Armoured Fighting Vehicles (AFV)
Australian Defence Forces (ADF)
Battle Management System (BMS)
Blue Force Tracking (BFT)
Capability Development Plan (CDP)
Central Command (CENTCOM)
Chemical, Biological, Radio-Nuclear (CBRN)
Civil-Military Co-Ordination (CMCO)
Close Air Support (CAS)
Coalition Agents Experiment (CoAX)
Coalition Shared Database (CSD)
Coalition Warfighter Interoperability Exercise/Demonstration (CWIX/D)
Collaborative Game for first Experiences in a Networked Environment (CAFFEINE)
Command and Battle Management (CBM)
Command and control (C2)
Command and Control Research Program (CCRP)
Command and Information System (CIS)
Command Control Communications Computers Intelligence Surveillance Reconnaissance (C4ISR)
Command Systems Interoperability (CSI)
Command, Control, Communications and Information (C3I)
Command, Control, Communications, Computers and Intelligence (C4I)
Commercial/Military-Off-The-Shelf Products (C/MOTS)
Common Operational Picture (COP)
Common Security and Defence Policy (CSDP)
Communications & Information Systems (CIS)
Comprehensive Approach (CA)
Concept Development & Experimentation (CD&E)
Cooperative Engagement Capability (CEC)
Customer Product Management (CPM)
Defence Equipment and Support (DE&S)
Defence Industry Strategy (DIS)
Defence Information Infrastructure (DII)
Defence Information Infrastructure Future Deployed (DII FD)
Defence Lines of Development (DLOD)
Defence Procurement Agency (DPA)
Defence Science and Technology Laboratory (DSTL)

Defence Scientific Advisory Council (DSAC)
Defence Technology Centre (DCC)
Defence Trade Security Initiative (DTSI)
Defense Capabilities Initiative (DCI)
Department of Defense (DoD)
Deputy Director of Equipment Capability for Command and Control Information Infrastructure (DDEC CCII)
Deputy Supreme Allied Commander Europe (DSACEUR)
Design, Build and Operate (DBO)
Digital Army Programme (DAP)
Digitisation Battlespace (Land) (DBL)
Doctrine, Organisation, Training, Material, Leadership Development, Personnel, and Facilities (DOTML-PF)
Effects-Based Operations (EBO)
Enterprise Architecture/Service Oriented Architecture (EA/SOA)
Equipment Capability Customer (ECC)
EU Network-Enabled Capability (EU NEC)
European Defence Agency (EDA)
European Security and Defence Policy (ESDP)
European Union (EU)
Forces Communications and Electronics Association (AFCEA)
Full Operational Capability (FOC)
Future Combat Systems (FCS)
Future Infantry Soldier Technology Programme (FIST)
Future Rapid Effects System (FRES)
Global Information Grid (GIG)
Global Positioning System (GPS)
Global Positioning System (GPS)
House of Commons (HoC)
Human Intelligence (HUMINT)
Improvised Explosive Devices (IED)
Industrie- und Anlagenbaugesellschaft (IABG)
Information Management (IM)
Information Technology (IT)
Initial Operational Capability (IOC)
Integrated Project Teams (IPT)
Integrated System Technologies (INSYTE)
Intelligence, Surveillance, Reconnaissance (ISR)
Interim Operations Support (IOS)
International Security Assistance Force (ISAF)
Israeli Defence Forces (IDF)
Italian Army Networking Programme (Forza NEC)
Joint Battlespace Digitisation (JBD)
Joint Capabilities Board (JCB)
Joint Center for International and Security Studies (JCISS)
Joint Command and Control Support Programme (JC2SP)
Joint Direct Attack Munition (JDAM)
Joint Effects Tactical Targeting System (JETTS)

Joint Experimentation, Transformation, and Concepts Division (JETCD)
Joint Functional Concepts (JFC)
Joint Integrating Concepts (JIC)
Joint Operational Command System (JOCS)
Joint Operational Picture (JOP)
Joint Operations Concept (JOpsC)
Joint Reference Information Database (JRMIC)
Joint Support Service (JSS)
Joint Surveillance, Target Attack Radar System (JSTAR)
Joint Tactical Fires (JTF)
Joint Vision 2010 (JV 2010)
Limited Objective Study on NEC (LOI NEC)
Long Term Vision for European Defence Capability and Capacity Needs (LTV)
Low Intensity Operations (LIO)
Manned Ground Vehicle (MGV)
Maritime/Command and Control Information System (M/CCIS)
Military Committee (MC)
Military Demonstration Exercise (DEMOEX)
Ministry of Defence (MoD)
Mission Capabilities Packages (MCP)
Multinational Experiment Series (MNE)
Multinational Interoperability Council (MIC)
National Audit Office (NAO)
NATO Architectural Framework (NAF)
NATO Command, Control and Communication Agency (NC3A)
NATO Communication and Information Systems Services Agency (NCSA)
NATO Consultation, Command, and Control Board (NC3B)
NATO Network-Enabled Capabilities (NNEC)
NATO Response Force (NRF)
Network Centric Operations Industry Consortium (NCOIC)
Network Enabled Capability (NEC)
Network-Based Defence (NBD)
Network-Based Operations (NBO)
Network-Centric Capability (NCC)
Network-Centric Collaborative Targeting (NCCT)
Network-Centric Operations (NCO)
Network-Centric Operations Conceptual Framework (NCO CF)
Network-Centric Operations Conceptual Framework (NCO CF)
Network-Centric Warfare (NCW)
Network-Enabled Air Capabilities (NEAC)
Network-Enabled Capabilities (NEC)
Network-Enabled Warfare (NEW)
Netzwerk-Orientierte Verfahren Austria (NOVA)
North Atlantic Council (NAC)
North Atlantic Treaty Organization (NATO)
Observe, Orient, Decide, Act (OODA)
Office of Force Transformation (OFT)
Operation Enduring Freedom (OEF)

Operation Iraqi Freedom (OIF)
Operational Mentoring and Liaison-Team (OMLT)
Permanent Joint Headquarters (PJHQ)
Prague Capabilities Commitment (PCC)
Precision Guided Munitions (PGM)
Private Finance Initiative (PFI)
Provincial Reconstruction Team (PRT)
Public Service Agreement (PSA)
Quick Reaction Force (QRF)
Rapid Prototyping, Development and Evaluation (RPDE)
Reconnaissance Airborne POD Tornado (RAPTOR)
Research and Development (R&D)
Revolution in Military Affairs (RMA)
Royal Air Force (RAF)
Science Applications International Corporation (SAIC)
Security and Defence Review (SDR)
Signal Intelligence (SIGINT)
Software-Defined Radio (SDR)
Standardization Agreements (STANAGs)
Strategic Defence and Security Review (SDSR)
Supreme Allied Commander Europe (SACEUR)
Supreme Allied Commander Transformation (SACT)
Supreme Headquarters Allied Powers Europe (SHAPE)
System für die Abbildende Aufklärung in der Tiefe des Einsatzraums (imaging surveillance system for the depth of the theatre) (SAATEG)
Transformation Coordinationá Group (TCG)
Transformation Planning Guidance (TPG)
U.S. Air Force Network Operations Command (AFNETOPS)
U.S. Cyber Command (USCYBERCOM)
U.S. Joint Forces Command (USJFCOM)
UK Forward Air Controller (FAC)
Ultra-High Frequency (UHF)
UN Security Council (UNSC)
United Kingdom (UK)
Unmanned Aerial Vehicle (UAV)
Urgent Operational Requirements (UOR)
Vernetzte Operationsführung (NetOpFü)

8 References

Arquilla, John (2008): *Worst Enemy*. Chicago, IL: Dee.

Arquilla, John/Ronfeldt, David (1997 [1993]): Cyberwar Is Coming! In: Arquilla, John/ Ronfeldt, David (Eds.) (1997 [1993]): In Athena's Camp. Preparing for Conflict in the Information Age. Santa Monica, CA: RAND Corporation, 23–60.

Arquilla, John/Ronfeldt, David (2003): Swarming – The Next Face of Battle. *Aviation Week & Space Technology*, 29 September. http://www.rand.org/commentary/100703AWST.html (accessed 05 May 2010).

Ackerman, Robert K. (2007): Sweden's Military Looks Outward. *SIGNAL Magazin (AFCEA)*, September. http://www.afcea.org/signal/articles/templates/SIGNAL_Article_Template.asp?articleid=1378&zoneid=41 (accessed 3 February 2013).

Ackerman, Robert K. (2008): Innovations Shape LandWarNet. *SIGNAL Magazin (AFCEA)*, August. www.afcea.org/content/?q=node/1661 (accessed 3 February 2013).

Ackerman, Robert K. (2009a): British Defense Information Technology Faces Uncertain Future. *SIGNAL Magazin (AFCEA)*, August. http://www.afcea.org/signal/articles/templates/Signal_Article_Template.asp?articleid=2027&zoneid=269 (accessed 20 June 2010).

Ackerman, Robert K. (2009b): German Government-Industry Relationship Sharply Defined. *SIGNAL Magazin (AFCEA)*, May. http://www.afcea.org/signal/articles/templates/Signal_Article_Template.asp?articleid=1935&zoneid=261 (accessed 20 June 2010).

Adamsky, Dima (2010): The Culture of Military Innovation: The Impact of Cultural Factors on the Revolution in Military Affairs in Russia, the US, and Israel. Stanford, CA: Stanford University Press.

Alberts, David S./Garstka, John J./Stein, Frederick (1999): Network Centric Warfare. Developing and Leveraging Information Superiority. Washington, D.C.: DoD Command and Control Research Program.

Alberts, David S./Garstka, John J./Hayes, Richard E./Signori, David A. (2001): Understanding Information Age Warfare. Washington, D.C.: CCRP Publication Series.

Alberts, David S./Hayes, Richard E. (2003): Power to the Edge: Command and Control in the Information Age. Washington, D.C.: CCRP Publication Series.

Alberts, David S./Hayes, Richard E. (2006 [2003]): Power to the Edge: Militärische Führung im Informationszeitalter. Bonn: German Ministry of Defence, Air Staff.

Allison, Graham T./Zelikow, Philip (1999): Essence of Decision: Explaining the Cuban Missile Crisis. New York: Longman.

Allsopp, David/Beautement, Patrick/Kirton, Michael/Bradshaw, Jeffrey M./Suri, Niranjan/Tate, Austin/Burstein, Mark (2003): The Coalition Agents Experiment: Network-Enabled Coalition Operations. *Journal of Defence Science,* 8: 3, 130–141.

Alston, Anthony (2003): Network-Enabled Capability – The Concept. *Journal of Defence Science*, 8: 3, 108–116.

Angerman, William S. (2004): Coming Full Circle With Boyd's OODA Loop Ideas: An Analysis of Innovation Diffusion and Evolution. Ohio: Air Force Institute Of Technology.

Argent-Hall, Dominic (2009): NEC in MOD Main Building: The Strength of Collaboration. In: Ministry of Defence (Ed.) (2009): NEC – Understanding Network Enabled Capability. London: Ministry of Defence, 67–69.

Avant, Deborah D. (1994): Political Institutions and Military Change: Lessons from Peripheral Wars. Ithaca, NY: Cornell University Press.

Baddeley, Adam (2006): Sweden Seeks Military Communications Flexibility. *SIGNAL Magazin (AFCEA)*, May. http://www.afcea.org/signal/articles/templates/SIGNAL_Article_Template.asp?articleid=1129&zoneid=7 (accessed 27 June 2010).

Baumgarten, Steven A. (1975): The Innovative Communicator in the Diffusion Process. *Journal of Marketing Research*, 12: 1, 12–18.

Baxter, Robert (2005): Ned Ludd Encounters Network-Enabled Capability. *RUSI Defence Systems*, 7: 3, 34–36.

Bennett, Colin J. (1991): How States Utilize Foreign Evidence. *Journal of Public Policy*, 11: 1, 31–54.

Betz, David J. (2006): The More You Know, the Less You Understand: The Problem with Information Warfare. *Journal of Strategic Studies*, 29: 3, 505–533.

Biddle, Stephen D. (2003): Afghanistan and the Future of Warfare. *Foreign Affairs*, 82: 2, 31–46.

Biddle, Stephen D. (2007): Speed Kills? Reassessing the Role of Speed, Precision, and Situation Awareness in the Fall of Saddam. *The Journal of Strategic Studies*, 30: 1, 3–46.

Binnendijk, Hans (2002): Introduction. In: Binnendijk, Hans (Ed.): Transforming America's Military. Washington, D.C.: Center for Technology and National Security Policy, National Defense University Press, xvii.

Bjerga, Kjell I./Haaland, Torunn L. (2010): Development of Military Doctrine: The Particular Case of Small States. *Journal of Strategic Studies*, 33: 4, 505–533.

Blackham, Jeremy (2000): Handling the Digitised Battlespace. *The RUSI Journal*, 145: 1, 33–37.

Blair, Tony (2010): A Journey. London: Hutchinson.

Blaker, James R. (2007): Transforming Military Force: The Legacy of Arthur Cebrowski and Network Centric Warfare. Westport, CN: Praeger Security International.

Book, Elizabeth G. (2002): Information Warfare Pioneers Take Top Pentagon Positions. *National Defence Magazine (online resource)*, January. http://www.nationaldefensemagazine.org/issues/2002/Jan/Information_Warfare.htm.

Boot, Max (2003): The New American Way of War. *Foreign Affairs,* 82: 4, 41–58.

Borchert, Heiko (2010): The Rocky Road to Networked and Effects-Based Expedionary Forces: Military Transformation in the Bundeswehr. In: Terriff, Terry/Osinga, Frans P. B./Farrell, Theo (Eds.): A Transformation Gap? American Innovations and European Military Change. Stanford, CA: Stanford University Press, 83–107.

Born, Hans/Beutler, Ingrid (2007): Between Legitimacy and Efficiency. A Comparative View on Democratic Accountability. In: Caforio, Giuseppe (Ed.): Social Sciences and the Military: An Interdisciplinary Overview. London – New York: Routledge, 261–286.

Boxenbaum, Eva/Jonsson, Stefan (2008): Isomorphism, Diffusion and Decouplin. In: Greenwood, Royston/Oliver, Christine/Sahlin, Kerstin/Suddaby, Roy (Eds.): The SAGE Handbook of Organizational Institutionalism. Los Angeles, CA – London: SAGE, 78–98.

Bracken, Paul (2002): Corporate Disasters. Some Lessons for Transformation. *Joint Forces Quarterly*, Autumn, 83–87.

Bredow, Wilfried von (2005): The Defence of National Territory: The German Experience. In: Edmunds, Timothy/Malešic, Marjan (Eds.): Defence Transformation in Europe: Evolving Military Roles. Amsterdam: IOS Press, 27–35.

Bredow, Wilfried von (2007): Militär und Demokratie in Deutschland. Eine Einführung. Wiesbaden: VS Verlag für Sozialwissenschaften.

British Army (2008): The Future Land Operational Concept. Shrivenham: The Joint Doctrine & Concepts Centre, Ministry of Defence.

British Chiefs of Staff (2001): British Defence Doctrine (JDP 0-01). 2nd Edition. Shrivenham: The Joint Doctrine & Concepts Centre, Ministry of Defence.

British Chiefs of Staff (2006): Information Management (Joint Doctrine Note 4/06). Shrivenham: The Joint Doctrine & Concepts Centre, Ministry of Defence.

British Chiefs of Staff (2008): British Defence Doctrine (JDP 0-01). 3rd Edition. Shrivenham: The Joint Doctrine & Concepts Centre, Ministry of Defence.

British Ministry of Defence (1996): Statement on the Defence Estimates 1996. London: Ministry of Defence.

British Ministry of Defence (1998): Strategic Defence Review. London: The Stationery Office.

British Ministry of Defence (2002): The Strategic Defence Review: The New Chapter. London: The Stationery Office.

British Ministry of Defence (2003a): Delivering Security in a Changing World. London: The Stationery Office.

British Ministry of Defence (2003b): Operations in Iraq: First Reflections (07/2003 C40). London: Ministry of Defence.

British Ministry of Defence (2005): Network Enabled Capability Handbook (Joint Services Publication 777). London: Ministry of Defence.

British Ministry of Defence (2009): NEC – Understanding Network Enabled Capability. London: Ministry of Defence.

British Ministry of Defence (2000/1-2007/8): Annual Report and Accounts (aka Progress Reports). London: Ministry of Defence. http://www.mod.uk/DefenceInternet/AboutDefence/CorporatePublications/AnnualReports/ (accessed 5 May 2010).

British Ministry of Defence (2011): MOD Information Strategy 2011. Better Informed, Better Defence. London: Ministry of Defence.

Brooks, Risa A. (2003): Making Military Might: Why Do States Fail and Succeed? A Review Essay. *International Security*, 28: 2, 149–191.

Brooks, Risa/Stanley, Elizabeth A. (2007): Creating Military Power: The Sources of Military Effectiveness. Stanford, CA: Stanford University Press.

Brown, Frederic J. (2001): Tactical Situational Awareness: The Human Challenge. In: Pfaltzgraff, Robert L., Jr./Schultz, Richard H., Jr. (Eds.): War in the Information Age. Washington, D.C. – London: Brassey's, 99–174.

Brown, Lawrence A. (1981): Innovation Diffusion: A New Perspective. London – New York: Methuen.

Bundesministerium der Verteidigung (2011): Die Neuausrichtung der Bundeswehr. Nationale Interessen wahren – Internationale Verantwortung übernehmen – Sicherheit gemeinsam gestalten. Berlin: BMVg.

Bush, George W. (1999): A Period of Consequences (Citadel Speech, 23 September 1999). Charleston, SC: The Citadel.

Bush, George W. (2001): Citadel Speech 2001 (Citadel Speech, 11 December 2001). Charleston, SC: The Citadel.

Cebrowski, Arthur K. (2002): An Interview with the Director, 28 October. *Transformation Trends*. http://www.cdi.org/mrp/tt-28oct02.pdf (accessed 15 October 2010).

Cebrowski, Arthur K. (2003a): Network-Centric Warfare: An Emerging Military Response to the Information Age. *Military Technology*, 27: 5, 16–22.

Cebrowski, Arthur K. (2003b): Speech by Arthur Cebrowski to the Network Centric Warfare Conference, 17 February. *Transformation Trends*. http://www.cdi.org/mrp/tt-17feb03.pdf (accessed 13 October 2010).

Cebrowski, Arthur K./Garstka, John J. (1998): Network-Centric Warfare: Its Origin and Future. *US Naval Institute Proceedings*, 124: 1, 28–35.

Chief of Staff of the German Air Force (2003): Konzeptionelle Position der Luftwaffe zur Vernetzten Operationsführung. Bonn: Inspekteur der Luftwaffe.

Chief of Staff of the German Army (2004): Positionspapier des Heeres zur Vernetzten Operationsführung. Bonn: Inspekteur des Heeres.

Chuter, Andrew (2004): U.K. Chops Forces: Troops, Tanks, Warships, Jets to Be Dumped. *Defense News*, 26 July. http://www.madisongov.net/news-articles.asp?article_type=2&article_link=072604_worde.txt&year=2004&title=U.K.%20Chops%20Forces (accessed 1 October 2010).

Cloud, David S./Schmitt, Eric (2006): More Retired Generals Call for Rumsfeld's Resignation. *New York Times,* 14 April. http://www.nytimes.com/2006/04/14/washington/14military.html?pagewanted=all&_r=0 (accessed 14 October 2010).

Codner, Michael (2005): Transformation: The Pursuit of Catalysis – A British View. In: Schreer, Benjamin/Whitlock, Eugene (Eds.): Divergent Perspectives on Military Transformation. Berlin: Stiftung Wissenschaft und Politik, 12–19.

Codner, Michael (2010): The Defence Review. Capability Questions for the New Government (*RUSI Working Paper*, No. 6). London: Royal United Services Institute for Defence & Security Studies (RUSI).

Cohen, Eliot A. (2004): Change and Transformation in Military Affairs. *Journal of Strategic Studies*, 27: 3, 395–407.

Cohen, William S. (1997): Report of the Quadrennial Defense Review. Washington, D.C.: Department of Defense.

Cohen, William S. (1999): Annual Defense Report. Washington, D.C.: Department of Defense.

Collmer, Sabine (2007): Information as a Key Resource: The Influence of RMA and Net-Centric Operations on the Transformation of the German Armed Forces. Garmisch-Partenkirchen: George C. Marshall Center for Security Studies.

Connaughton, Richard M. (2000): Organizing British Joint Rapid Reaction Forces. *Joint Forces Quarterly*, Autumn, 87-94.

Corum, James S. (1994): A Clash of Military Cultures: German and French Approaches to Technology between the World Wars. Maxwell, Montgomery (Alabama): USAF Air University. http://www.au.af.mil/au/awc/awcgate/saas/corum.pdf (accessed 23 April 2011).

Coward, Gary (2009): Understanding Joint Action. In: Ministry of Defence (Ed.): NEC – Understanding Network Enabled Capability. London: Ministry of Defence, 24.

Creveld, Martin L. van (1991): The Transformation of War. New York: The Free Press.

Czelusta, Mark G. (2008): Business as Usual: An Assessment of Donald Rumsfeld's Transformation Vision and Transformation's Prospects for the Future. Garmisch-Partenkirchen: George C. Marshall Center for Security Studies.

Deephouse, David L./Suchman, Mark C. (2008): Legitimacy in Organizational Institutionalism. In: Greenwood, Royston/Oliver, Christine/Sahlin, Kerstin/Suddaby, Roy (Eds.): The SAGE Handbook of Organizational Institutionalism. Los Angeles – London: SAGE, 49–77.

Defence Science and Technology Laboratory (2002–2010): Annual Report. http://www.dstl.gov.uk/corporatepublications (accessed 2 February 2011).

Defence Science and Technology Laboratory (2003): NEC Outline Concept (Dstl/IMD/SOS/500/2). Salisbury: DSTL.

Della Porta, Donatella/Keating, Michael (2008): Approaches and Methodologies in the Social Sciences: A Pluralist Perspective. Cambridge, MA – New York: Cambridge University Press.

Demchak, Chris (2003): Creating the Enemy. Global Diffusion of the Technology-Based Military Model. In: Goldman, Emily O./Eliason, Leslie C. (Eds.): The Diffusion of Military Technology and Ideas. Stanford, CA: Stanford University Press, 307–347.

DiMaggio, Paul J. (1988): Interest and Agency in Institutional Theory. In: Zucker, Lynne G. (Ed.): Institutional Patterns and Organizations: Culture and Environment. Cambridge: Ballinger, 3–21.

DiMaggio, Paul J./Powell, Walter W. (1991 [1984]): The Iron Cage Revisited: Institutional Isomorphism and Collective Rationality in Organizational Fields. In: Powell, Walter W./DiMaggio, Paul J. (Eds.): The New Institutionalism in Organizational Analysis. Chicago, IL: University of Chicago Press, 63–82.

Dombrowski, Peter J./Gholz, Eugene (2006): Buying Military Transformation: Technological Innovation and the Defense Industry. New York: Columbia University Press.

Downs, Edward (2009): ISTAR Operational and System Architectures. In: Ministry of Defence (Ed.): NEC–Understanding Network Enabled Capability. London: Ministry of Defence, 46–48.

Dumbrell, John (2009): The US-UK Special Relationship: Taking the 21st-Century Temperature. *British Journal of Politics and International Relations*, 11: 1, 64–78.

Dyson, Tom (2005): German Military Reform 1998–2004: The Triumph of Domestic Constraint over International Opportunity. *European Security*, 14: 3, 361–386.

Dyson, Tom (2008): Convergence and Divergence in Post-Cold War British, French, and German Military Reforms. Between International Structure and Executive Autonomy. *Security Studies*, 17: 4, 725–774.

Dyson, Tom (2013): German Defence Politics: A View from Abroad. In: Wiesner, I. (Ed.): Understanding German Defence. Baden-Baden: Nomos (in print).

Easton, David (1965): A Framework for Political Analysis. Englewood Cliffs, N.J.: Prentice-Hall.

Ebbutt, Giles (2006): Flaws in the System: Modern Operations Test the Theory of Network Centricity. *International Defence Review*, 39: 7, 57.

Ebbutt, Giles (2008): UK Forces in Afghanistan Begin to See NEC Dawn. *International Defence Review*, 28 May. http://search.janes.com (accessed 9 June 2009).

Egnell, Robert (2006): Explaining US and British Performance in Complex Expeditionary Operations: The Civil-Military Dimension. *Journal of Strategic Studies*, 29: 6, 1041–1075.

Eliason, Leslie C./Goldman, Emily O. (2003): Introduction. Theoretical and Comparative Perspectives on Innovation and Diffusion. In: Goldman, Emily O./ Eliason, Leslie C. (Eds.): The Diffusion of Military Technology and Ideas. Stanford, CA: Stanford University Press, 1–32.

Eriksson, Johan/Giacomello, Giampiero (2006): The Information Revolution, Security, and International Relations: (IR) Relevant Theory? *International Political Science Review*, 27: 3, 221–244.

Eshel, Tamir (1991): The Most Successful Air Campaign Ever? *Military Technology*, 4, 36–44.

Eyre, Dana P./Suchman, Mark C. (1996): Status, Norms, and the Proliferation of Conventional Weapons: An Institutional Theory Approach. In: Katzenstein, Peter J. (Ed.): The Culture of National Security: Norms and Identity in World Politics. New York: Columbia University Press, 80–113.

Farrell, Theo (1998): Review: Culture and Military Power. *Review of International Studies*, 24: 3, 407–416.

Farrell, Theo (2001): Transnational Norms and Military Development: Constructing Ireland's Professional Army. *European Journal of International Relations*, 7: 1, 63–102.

Farrell, Theo (2008): The Dynamics of British Military Transformation. *International Affairs*, 84: 4, 777–807.

Farrell, Theo/Bird, Tim (2010): Innovating within Cost and Cultural Constraints: The British Approach to Military Transformation. In: Terriff, Terry/Osinga, Frans P. B./Farrell, Theo (Eds.): A Transformation Gap? American Innovations and European Military Change. Stanford, CA: Stanford University Press, 36–58.

Farrell, Theo/Terriff, Terry (2002): The Sources of Military Change. In: Farrell, Theo/ Terriff,Terry (Eds.): The Sources of Military Change: Culture, Politics, Technology. Boulder, CO: Lynne Rienner Publishers, 3–20.

Farrell, Theo/Terriff, Terry (2010): Military Transformation in NATO: A Framework for Analysis. In: Terriff, Terry/Osinga, Frans P. B./Farrell, Theo (Eds.): A Transformation Gap? American Innovations and European Military Change. Stanford, CA: Stanford University Press, 1–13.

Fearon, James D./Wendt, Alexander (2002): Rationalism v. Constructivism: A Sceptical View. In: Carlsnaes, Walter/Risse-Kappen, Thomas/Simmons, Beth A. (Eds.): Handbook of International Relations. London – Thousand Oaks, CA: SAGE Publications, 52–72.

Ferbrache, David (2003): Network Enabled Capability: Concepts and Delivery. *Journal of Defence Science*, 8: 3, 104–107.

Fontenot, Gregory/Degen, E. J./Tohn, David/United States. Army. Operation Iraqi Freedom Study Group. (2005): On Point: The United States Army in Operation Iraqi Freedom. Annapolis, Md.: Naval Institute Press.

Fordham, Benjamin O. (2004): A very Sharp Sword: The Influence of Military Capabilities on American Decisions to Use Force. *Journal of Conflict Resolution*, 48: 5, 632–656.

Forster, Michael (2009): Beschaffungen 2009: 4711. *geopowers*, 21 January. http://www.geopowers.com/Machte/Deutschland/Rustung/Rustung_2009 (accessed 10 December 2009).
Fuller, John F. Ch. (1926): The Foundations of the Science of War. London: Hutchinson & Co. (Electronic Reprint: Fort Leavenworth, KS, U.S. Army Command and General Staff Press, 1993). http://www.usacac.army.mil/cac2/cgsc/carl/download/csipubs/FoundationsofScienceofWar.pdf (accessed 3 February 2013).
Fulton, Rob (2003): Network Enabled Capability (Foreword). *Journal of Defence Science*, 8: 3, 103.
Garamone, Jim (2010): Gates Puts Meat on Bones of Department Efficiencies Initiative. *American Forces Press Service*, 9 August. http://www.defense.gov//News/NewsArticle.aspx?ID=60348.
Garstka, John J. (2000): Network Centric Warfare: An Overview of Emerging Theory. *Phalanx Online*, 33: 4, 1–33.
Garstka, John J. (2003): Network Centric Operations Conceptual Framework Version 1.0. Prepared by: Evidence Based Research, Inc. 1595 Spring Hill Rd Suite 250 Vienna, VA. http://www.dtic.mil (accessed 23 April 2011).
Garstka, John J./Alberts, David S. (2004): Network Centric Operations Conceptual Framework Version 2.0. Evidence Based Research, Inc.: Vienna, VA. http://www.dtic.mil (accessed 23 April 2011).
Garstka, John J./Holloman, Kimberly/Kiamie, Bud/Lewis, Joseph (2006): Network Centric Operations (NCO) Case Study. Coalition Operations in Operation IRAQI FREEDOM (OIF): A U.K. Perspective of FBCB2/Blue Force Tracker (BFT). Washington, D.C.: Department of Defense, Office of Force Transformation.
Gates, Robert (2010): Report of the Quadrennial Defense Review. Washington, D.C.: Department of Defense.
George, Alexander L./Bennett, Andrew (2004): Case Studies and Theory Development in the Social Sciences. Cambridge, MA – London: MIT Press.
German Ministry of Defence (2003): Weisung für die Weiterentwicklung der Bundeswehr. Bonn – Berlin: German Ministry of Defence.
German Ministry of Defence (2004): Konzeption der Bundeswehr. Bonn: FüS VI 2.
German Ministry of Defence (2005): Grundsätze für Aufgabenzuordnung, Organisation und Verfahren im Bereich der militärischen Spitzengliederung (Berliner Erlass). Berlin: German Ministry of Defence.
German Ministry of Defence (2006a): Teilkonzeption Vernetzte Operationsführung (TK NetOpFü). Bonn: FüS VI 2.
German Ministry of Defence (2006b): White Paper 2006 on German Security Policy and the Future of the Bundeswehr. Bonn – Berlin: German Ministry of Defence.
German Ministry of Defence (2012): Grundsätze für die Spitzengliederung, Unterstellungsverhältnisse und Führungsorganisation im Bundesministerium der Verteidigung und der Bundeswehr (Dresdener Erlass). Berlin: German Ministry of Defence.
Goldman, Emily O. (2003): Receptivity to Revolution. Carrier Air Power in Peace and War. In: Goldman, Emily O./Eliason, Leslie C. (Eds.): The Diffusion of Military Technology and Ideas. Stanford, CA: Stanford University Press, 267–303.

Goldman, Emily O./Ross, Andrew L. (2003): The Diffusion of Military Technology and Ideas – Theory and Practice. In: Goldman, Emily O./Eliason, Leslie C. (Eds.): The Diffusion of Military Technology and Ideas. Stanford, CA: Stanford University Press, 371–403.

Gormley, Dennis M./Hart, Douglas M. (2000): Extending Network-Centric Warfare to Coalition Crisis Management and Assessment. *The RUSI Journal*, 145: 2, 67–72.

Graham, Bradley (2005): Pentagon Prepares to Rethink Focus on Conventional Warfare. *The Washington Post,* 26 January.

Gray, Colin S. (2005): Transformation and Strategic Surprise. Carlisle, PA: Strategic Studies Institute of the U.S. Army War College. http://www.Strategicstudiesinstitute.army.mil/pdffiles/PUB602.pdf (accessed 10 November 2009).

Gray, Colin S. (2006a): Irregular Enemies and the Essence of Strategy: Can the American Way of War Adapt? Carlisle, PA: Strategic Studies Institute of the U.S. Army War College. http://www.strategicstudiesinstitute.army.mil/pdffiles/PUB 640.pdf (accessed 10 November 2009).

Gray, Colin S. (2006b): Reorganizing and Understanding Revolutionary Change in Warfare: The Sovereignty of Context. Carlisle, PA: Strategic Studies Institute of the U.S. Army War College. http://www.strategicstudiesinstitute.army.mil/pdf files/PUB640.pdf (accessed 10 November 2009).

Gray, Colin S. (2006c): Technology as a Dynamic of Defence Transformation. *Defence Studies*, 6: 1, 26–51.

Greenwood, Royston/Oliver, Christine/Sahlin, Kerstin/Suddaby, Roy (2008): Introduction. In: Greenwood, Royston/Oliver, Christine/Sahlin, Kerstin/Suddaby, Roy (Eds.): The SAGE Handbook of Organizational Institutionalism. Los Angeles – London: SAGE, 1–46.

Griffin, Stuart/Whetham, David (2007): Case Study 7 – Iraq. In: Office of Force Transformation (Ed.): Network Centric Operations (NCO). Case Study: The British Approach to Low-Intensity Operations: Part II. Washington, D.C.: Office of Force Transformation.

Grissom, Adam (2006): The Future of Military Innovation Studies. *Journal of Strategic Studies*, 29: 5, 905–934.

Hammes, Thomas X (2004): The Sling and the Stone. St. Paul (U.S.): Zenith Press.

Harnisch, Sebastian (2005): Deutsche Außenpolitik auf dem Prüfstand. Die Kultur der Zurückhaltung und die Debatte über nationale Interessen. *Reader Sicherheitspolitik, Ergänzungslieferung*, 3: 5, 10–24.

Harnisch, Sebastian/Longhurst, Kerry (2006): Understanding Germany: The Limits of 'Normalization' and the Prevalence of Strategic Culture. In: Taberner, Stuart/Cooke, Paul (Eds.): German Culture, Politics, and Literature into the Twenty-First Century: Beyond Normalization. Rochester, NY: Camden House, 49–60.

Hawk, Jeff (2005): Technology Tracks Casualties, Assets. *SIGNAL Magazin (AFCEA)*, November. http://www.afcea.org/signal/articles/templates/SIGNAL_Article_Template.asp?articleid=1043&zoneid=168 (accessed 28 June 2010).

Her Majesty's Government (2010): The Strategic Defence and Security Review. London: The Stationery Office/Tso.

Héritier, Adrienne (2007): Explaining Institutional Change in Europe. Oxford – New York: Oxford University Press.

Heuser, Beatrice (1998): Nuclear Mentalities? Strategies and Beliefs in Britain, France and the FRG. Basingstoke: Macmillan.

Hilpert, Carolin (2010): Deutsche Strategiefähigkeit im 21. Jahrhundert: Hintergrund: Kultur der Zurückhaltung. */e-politik.de/e.V. Onlinemagazin für Politik, Gesellschaft und Politikwissenschaft*, 6 July. http://www.e-politik.de/lesen/artikel/2010/hintergrund-kultur-der-zuruckhaltung/print/ (accessed 10 November 2010).

Hirschman, Albert O. (1970): Exit, Voice, Loyalty. Responses to Decline in Firms, Organizations and States. Cambridge, MA – London: Harvard University Press.

Hofstede, Geert H./Hofstede, Gert Jan (2005): Cultures and organizations : software of the mind. New York: McGraw-Hill.

Holt, John (2003): NEC Social and Organizational Factors. *Journal of Defence Science*, 8: 3, 142–151.

Honekamp, Wilfried (Ed.) (2008): Concept Development & Experimentation. Erfahrungen aus der praktischen Anwendung der Methode zur Transformation von Streitkräften. Remscheid: Re Di Roma Verlag.

Horowitz, Michael C. (2010a): The Diffusion of Military Power. Causes and Consequences for International Politics. Princeton, NJ: Princeton University Press.

Horowitz, Michael C. (2010b): Nonstate Actors and the Diffusion of Innovations: The Case of Suicide Terrorism. *International Organization*, 64: 1, 33–64.

Hughes, Kent/Werwatz, Axel (2006): Innovation in the United States and Germany (*AICGS Policy Report*). Washington, D.C.: AICGS. http://www.aicgs.org/documents/polrep26.pdf (accessed 12 December 2010).

Huntington, Samuel P. (1957): The Soldier and the State. The Theory and Politics of Civil-Military Relations. Cambridge; MA: Belknap Press of Harvard University Press.

Iraq Study Group (2006): The Iraq Study Group Report. Washington, D.C.: United States Institute of Peace. http://www.usip.org/isg/iraq_study_group_report/report/1206/index.html (accessed 3 February 2013).

Jackson, Mike (2007): Soldier: The Autobiography of General Sir Mike Jackson. London: Bantam.

Jacoby, Wade (2004): The Enlargement of the European Union and NATO: Ordering from the Menu in Central Europe. Cambridge – New York: Cambridge University Press.

James, Andrew/Meyer, David (2009): The UK's NEC Strategy. In: Ministry of Defence (Ed.): NEC – Understanding Network Enabled Capability. London: Ministry of Defence, 12–14.

Joint Doctrine and Concepts Centre (2001): The UK Joint Vision. Shrivenham: JDCC, Ministry of Defence.

Joint Doctrine and Concepts Centre (2003): The UK Joint High Level Operational Concept v. 1.3. Shrivenham: Joint Doctrine and Concepts Centre.

Jones, Llyr (2008): Niteworks: Establishing a UK Strategic Asset. *RUSI Defence Systems*, 11: 2, 100–103.

Jung, Franz J. (2006): Rede des Bundesministers der Verteidigung anlässlich des Besuchs beim Zentrum für Transformation der Bundeswehr am 22. Mai 2006 in Strausberg. Berlin: BMVg.

Junio, Timothy J. (2009): Military History and Fourth Generation Warfare. *Journal of Strategic Studies:* 32: 2, 243–269.

Kagan, Frederick W. (2006): Finding the Target: The Transformation of American Military Policy. New York: Encounter Books.

Katzenstein, Peter J. (1996): Introduction. In: Katzenstein, Peter J. (Ed.): The Culture of National Security: Norms and Identity in World Politics. New York: Columbia University Press, 1–32.

Kaufman, Alfred (2002): Be Careful What You Wish for: The Dangers of Fighting with a Network-Centric Military. *Journal of Battlefield Technology*, 5: 2, 20–23.

Kaufman, Alfred (2005): Caught in the Network. How the Doctrine of Network-Centric Warfare Allows Technology to Dictate Military Strategy. *Armed Forces Journal*, 142: 7, 20–22.

Kaufman, Alfred/Welch, Larry D. (2004): Curbing Innovation: How Command Technology Limits Network Centric Warfare. Canberra: Argos.

Kemp, Damian (2003): UK/US Interoperability on Track. *Jane's Defence Weekly*, 8 October. http://search.janes.com (accessed 23 April 2011).

Kenyon, Henry S. (2007a): Future Combat Systems Begin to Deploy. *SIGNAL Magazin (AFCEA),* November. http://www.afcea.org/signal/articles/templates/Signal_Article_Template.asp?articleid=1414&zoneid=219 (accessed 20 September 2009).

Kenyon, Henry S. (2007b): Sweden Strives for Collaborative Operations. *SIGNAL* Magazin (AFCEA), September. http://www.afcea.org/signal/articles/templates/SIGNAL_Article_Template.asp?articleid=1380&zoneid=47 (accessed 20 September 2009).

Kenyon, Henry S. (2008): Training Vital to Network Defense. *SIGNAL Magazin (AFCEA)*, August. http://www.afcea.org/signal/articles/templates/Signal_Article_Template.asp?articleid=1665&zoneid=238 (accessed 20 September 2009).

Keymer, Eleanor (2009): Jane's World Armies. Virginia: Jane's Information Group Inc.

Kincaid, Bill (2008): Changing the Dinosaur's Spots: The Battle to Reform UK Defence Acquisition. London: RUSI.

Klos, Dietmar (2006): Vernetzte Operationsführung des Heeres im internationalen Kontext. *Europäische Sicherheit*, 55: 5, 54–61.

Klos, Dietmar/Möllers, Heiner/Stockfisch, Dieter (2013): Die Teilstreitkräfte. In: Wiesner, Ina (Ed.): Deutsche Verteidigungspolitik. Baden-Baden: Nomos (in print).

Knittlmaier, Hans-Jürgen (2010): Erweiterung der Aufklärungsfähigkeit der Luftwaffe. *Strategie & Technik*, June, 19–25.

Koydl, Wolfgang (2010): Mehr EU, weniger USA – Neue britische Regierung ändert außenpolitischen Kurs. *Süddeutsche Zeitung*, 3 July. http://www.sueddeutsche.de/politik/grossbritannien-mehr-eu-weniger-usa-1.969297 (accessed 16 December 2010).

Krepinevich, Andrew F. (1994): Cavalry to Computer. The Pattern of Military Revolution. *The National Interest*, Fall, 30–42.

Krepinevich, Andrew F. (2002 [1992]): The Military-Technical Revolution. A Preliminary Assessment. Washington, D.C.: Center for Strategic and Budgetary Assessments.

Krepinevich, Andrew F. (2002): Defense Transformation. Washington, D.C.: CSBA, 4 September. http:// www.csbaonline.org/4Publications/PubLibrary/T. 20020409. Defense_Transforma/T.20020409.Defense_Transforma.php (accessed 14 October 2010).

Kümmel, Gerhard (2003): The Winds of Change: The Transition from Armed Forces for Peace to New Missions for the Bundeswehr and its Impact on Civil-Military Relations. *Journal of Strategic Studies*, 26: 2, 7–28.

Kupchan, Charles (2010): Britain Is no longer America's Bridge to Europe. *Financial Times*, 1 June 2010. http://www.ft.com/cms/s/0/dc5cf340-6dcb-11df-b5c9-00144feabdc0.html (accessed 1 July 2010).

Lake, Darren (2003): UK Clarifies Progress on Network-Centric Warfare. *Jane's Defence Weekly*, 7 March. http://search.janes.com (accessed 23 April 2011).

Lange, Heinrich (2008): Vernetzte Operationsführung – Übergreifendes Prinzip der Transformation. *Wehrtechnischer Report*, 6, 3.

Laupert, Stefan (2008): Vernetzte Operationsführung im Transformationsprozess. *Wehrtechnischer Report*, 6, 6–8.

Lawlor, Maryann (2006): Restructuring Boosts Navy Information Sharing. *SIGNAL* Magazin (AFCEA), December. http://www.afcea.org/signal/articles/templates/Signal_Article_Template.asp?articleid=1228&zoneid=195 (accessed 27 Juni 2010).

Lawlor, Maryann (2007): Engineering Network-Centric Warfare. SIGNAL Magazin (AFCEA), August. http://www.afcea.org/signal/articles/templates/SIGNAL_Article_Template.asp?articleid=1360&zoneid=40 (accessed 22 June 2010).

Le Fevre, Graham/Thornton, John (2003): The Case for Manned Reconnaissance. In: Potts, David (Ed.): The Big Issue: Command and Combat in the Information Age. Washington, D.C.: CCRP Publication Series: 189–198.

Levy, Jack S. (2008): Case Studies: Types, Designs, and Logics of Inference. *Conflict Management and Peace Science*, 25: 1, 1–18.

Lock-Pullan, Richard (2005): How to Rethink War: Conceptual Innovation and AirLand Battle Doctrine. *Journal of Strategic Studies*, 28: 4, 679–702.

Longhurst, Kerry (2005a): Endeavors to Restructure the Bundeswehr: The Reform of the German Armed Forces 1990–2003. *Defense & Security Analysis*, 21: 1, 21–36.

Longhurst, Kerry (2005b): Germany and the Use of Force. Manchester: Manchester University Press.

Longhurst, Kerry/Miskimmon, Alister (2007): Same Challenges, Diverging Responses: Germany, the UK and European Security. *German Politics*, 16: 1, 79–94.

Lynn, John A. (2001): Reflections on the History and Theory of Military Innovation and Diffusion. In: Elman, Colin/Elman, Miriam F. (Eds.): Bridges and Boundaries: Historians, Political Scientists, and the Study of International Relations. Cambridge, MA: MIT Press: 359–382.

Mahaffey, John/Skaar, Trond (2005): Dissemination of ISR Data with Networked-Enabled Capabilities in a Coalition Environment. *Military Technology*, 29: 8, 61–67.

Mahajan, Vijay/Muller, Eitan/Srivastava, Rajendra K. (1990): Determination of Adopter Categories by Using Innovation Diffusion Models. *Journal of Marketing Research*, 27: 1, 37–50.

Mahnken, Thomas G. (2000): War and Culture in the Information Age. *Strategic Review*, Winter, 40–46.

Mahnken, Thomas G. (2008): Technology and the American Way of War. New York: Columbia University Press.

Majone, Giandomenico (1991): Cross-National Sources of Regulatory Policymaking in Europe and the United States. *Journal of Public Policy*, 11: 1, 79–106.

Marahrens, Sönke (2006): Entwicklung Vernetzter Operationsführung in der Luftwaffe. *Europäische Sicherheit*, 55: 6, 26–30.

March, James G./Olsen, Johan P. (1995): Democratic Governance. New York: Free Press.
March, James G./Olsen, Johan P. (2004): The Logic of Appropriateness (*Arena Working Paper*, No. 9). Oslo: Arena. http://www.sv.uio.no/arena/english/research/publications/arena-publications/workingpapers/working-papers2004/wp04_9.pdf (accessed 3 February 2013).
March, James G./Simon, Herbert A./Guetzkow, Harold Steere (1993): Organizations. Cambridge, MA: Blackwell.
Maull, Hanns W. (1990): Germany and Japan: The New Civilian Powers. *Foreign Affairs*, 69: 5, 91–116.
McCarthy, Jim/Arthur, Stan/DeMarines, Vic/Gold, Ted/Graham, Bill/Hartzog, Bill/Kaminski, Dr. Paul/Mundy, Carl/Studeman, Bill/Welch, Larry (2001): Transformation Study Report: Transforming Military Operational Capabilities (Secretary of Defense Report). Washington, D.C.: Department of Defense.
Meiter, John S. (2006): Network Enabled Capability: A Theory Desperately in Need of Doctrine. *Defence Studies*, 6: 2, 189–214.
Merriam-Webster Inc. (1998): Merriam-Webster's Collegiate Dictionary. Springfield, MA: Merriam-Webster.
Mey, Holger H./Krüger, Michael K.-D. (2003): Vernetzt zum Erfolg? 'Network-Centric Warfare' – Zur Bedeutung für die Bundeswehr. Eine Studie in Zusammenarbeit mit der EADS Company. Frankfurt am Main – Bonn: Report Verlag.
Meyer, John W./Rowan, Brian (1991): Institutionalized Organizations: Formal Structure as Myth and Ceremony. In: Powell, Walter W./DiMaggio, Paul J. (Eds.): The New Institutionalism in Organizational Analysis. Chicago, IL: University of Chicago Press, 41–62.
Millett, Allan R./Murray, Williamson/Watman, Kenneth H. (1986): The Effectiveness of Military Organizations. *International Security,* 11: 1, 37–71.
Millett, Allan Reed/Murray, Williamson (1988): Military Effectiveness. Boston: Unwin Hyman.
Mjøset, Lars (2001): Theory: Conceptions in the Social Sciences. In: Smelser, Neil J./Baltes, Paul B. (Eds.): International Encyclopedia of the Social and Behavioral Sciences. Oxford – New York: Elsevier, 15641–15647.
Moskos, Charles C./Williams, John Allen/Segal, David R. (Eds.) (2000): The Postmodern Military: Armed Forces after the Cold War. New York: Oxford University Press.
Murphy, D. (2005): Network Enabled Operations in Operation Iraqi Freedom. Carlisle, PA: Center for Strategic Leadership.
Murray, Williamson (1997): Thinking About Military Revolutions. *Joint Forces Quarterly*, Summer, 69–76.
Murray, Williamson (2003): Lessons Learned and Not Learned: The Gulf War in Retrospect. In: Brands, Henry W. (Ed.): The Use of Force after the Cold War. College Station, TX – London: Texas A&M University Press, 93–110.
Murray, Williamson/Millett, Allan Reed (1996): Military Innovation in the Interwar Period. Cambridge, MA – New York: Cambridge University Press.
Murray, Williamson/Scales, Robert H. (2003): The Iraq War: A Military History. Cambridge, MA: Belknap Press of Harvard University Press.
Nagl, John A. (2005 [2002]): Learning to Eat Soup with a Knife: Counterinsurgency Lessons from Malaya and Vietnam. Chicago, IL: University of Chicago Press.

Neureuther, Jörg (2008): Erfahrungen im multinationalen Rahmen am Beispiel der Multinationalen Experimentserie. In: Honekamp, Wilfried (Ed.): Concept Development & Experimentation. Erfahrungen aus der praktischen Anwendung der Methode zur Transformation von Streitkräften. Remscheid: Re Di Roma Verlag, 134–142.

North, Douglass Cecil (1990): Institutions, Institutional Change, and Economic Performance. Cambridge, MA – New York: Cambridge University Press.

Nye, Joseph S., Jr./Owens, William A. (1996): America's Information Edge. *Foreign Affairs*, 75: 2, 20–36.

O'Hanlon, Michael (2002): Rumsfeld's Defence Vision. *Survival: Global Politics and Strategy*, 44: 2, 103–117.

O'Hanlon, Michael E. (2009): The Science of War: Defense Budgeting, Military Technology, Logistics, and Combat Outcomes. Princeton, N.J.: Princeton University Press.

O'Rourke, R. (2005): Navy Network-Centric Warfare Concept: Key Programs and Issues for Congress. Washington, D.C.: Library of Congress.

O'Neil, William D. (2002): The Naval Services: Network-Centric Warfare. In: Binnendijk, Hans (Ed.): Transforming America's Military. Washington, D.C.: Center for Technology and National Security Policy, National Defense University Press, 129–158.

Office of Force Transformation (2003): Military Transformation: A Strategic Approach. Washington, D.C.: Office of Force Transformation.

Oliver, Christine (1991): Strategic Responses to Institutional Processes. *The Academy of Management Review*, 16: 1, 145–179.

Osinga, Frans (2010): The Rise of Military Transformation. In: Terriff, Terry/Osinga, Frans P. B./Farrell, Theo (Eds.): A Transformation Gap? American Innovations and European Military Change. Stanford, CA: Stanford University Press, 14–34.

Owens, William A. (2001): Lifting the Fog of War. Baltimore, MD: Johns Hopkins University Press.

Perry, William J. (1991): Desert Storm and Deterrence. *Foreign Affairs*, 70: 4, 66–82.

Pfeffer, Jeffrey/Salancik, Gerald R. (1978): The External Control of Organizations: A Resource Dependence Perspective. Stanford, CA: Stanford Business Books.

Posen, Barry (1984): The Sources of Military Doctrine: France, Britain, and Germany between the World Wars. Ithaca, NY: Cornell University Press.

Potts, David (2003): The Big Issue: Command and Combat in the Information Age. Washington, D.C.: CCRP Publication Series.

Powell, Walter W. (1991): Expanding the Scope of Institutional Analysis. In: Powell, Walter W./DiMaggio, Paul J. (Eds.): The New Institutionalism in Organizational Analysis. Chicago, IL: University of Chicago Press, 183–203.

Price, Richard/Tannenwald, Nina (1996): Norms and Deterrence: The Nuclear and Chemical Weapons Taboos. In: Katzenstein, Peter J. (Ed.): The Culture of National Security: Norms and Identity in World Politics. New York: Columbia University Press: 114–152.

Quintana, Elizabeth (2007): Is NEC Dead? (*RUSI Paper*). London: RUSI. http://www.rusi.org/downloads/ assets/NEC2007.pdf (accessed 12 June 2009).

Risen, Clay (2006): War-Mart. The Danger of Generals-as-CEOs. *The New Republic*. http://www.tnr.com/print/article/the-danger-generals-ceos (accessed 12 June 2009).

Rogers, Everett M. (1960): Social Change in Rural Society. A Textbook in Rural Sociology. New York: Appleton-Century-Crofts.
Rogers, Everett M. (1968): Supplement to Bibliography on the Diffusion of Innovations (*Diffusion of Innovations Research Report*, No. 6a). East Lansing, MI: Department of Communication, Michigan State University.
Rogers, Everett M. (1995): Diffusion of Innovations. 4th Edition. New York: Free Press.
Rogers, Everett M. (2003): Diffusion of Innovations. 5th Revised Edition. New York: Free Press.
Rogers, Everett M./Williams, Linda/West, Rhonda B. (1967): Bibliography on the Diffusion of Innovations. East Lansing, MI: Department of Communication, Michigan State University.
Rosen, Stephen P. (1991): Winning the Next War: Innovation and the Modern Military. Ithaca, NY – London: Cornell University Press.
Rosen, Stephen P. (1995): Military Effectiveness: Why Society Matters. *International Security*, 19: 4, 5–31.
Rosen, Stephen P. (2010): The Impact of the Office of Net Assessment on the American Military in the Matter of the Revolution in Military Affairs. *Journal of Strategic Studies*, 33: 4, 469–482.
Roxborough, Ian (2002): From Revolution to Transformation: The State of the Field. *Joint Forces Quarterly*, Autumn, 68–78.
Royal Air Force (2006): The Royal Air Force Strategy: Agile, Adaptable, Capable. London: Ministry of Defence (Directorate of Air Staff).
Royal Navy (2004): Future Maritime Operational Concept. NAVB/P (04). Shrivenham: Development, Concepts and Doctrine Centre, Ministry of Defence.
Rumsfeld, Donald (2001a): Prepared Testimony of U.S. Secretary of Defense Donald H. Rumsfeld Washington, D.C.: Senate Armed Services Committee, Thursday, 21 June.
Rumsfeld, Donald (2001b): Report of the Quadrennial Defense Review. Washington D.C.: Department of Defense.
Rumsfeld, Donald (2002a): 21st Century Transformation of U.S. Armed Forces. Speech given at the National Defense University, Fort McNair, Washington, D.C., Thursday, 31 January. Washington, D.C.: Office of the Assistant Secretary of Defense (Public Affairs).
Rumsfeld, Donald (2002b): Transforming the Military. *Foreign Affairs*, 81: 2, 20–32.
Rumsfeld, Donald (2003): Prepared Testimony, 9 July. Washington, D.C.: Senate Armed Services Committee.
Rumsfeld, Donald (2006): Report of the Quadrennial Defense Review. Washington D.C.: Department of Defense.
Sarvary, Miklos/Parker, Philip M./Dekimpe, Marnik G. (2000): Global Diffusion of Technological Innovations: A Coupled-Hazard Approach. *Journal of Marketing Research*, 37: 1, 47–59.
Scarborough, Rowan (2004): Rumsfeld's War. The Untold Story of America's Anti-Terrorist Commander. Washington, D.C.: Regnery Publishing, Inc.
Schäfer, Sebastian (2005): Vernetzte Operationsführung (NetOpFü) – Eine Einführung. Cologne-Wahn: Zentrum für Weiterentwicklung der Luftwaffe.
Scherz, Reimar (2003): Speech Held by Brigadier General Scherz on the 2nd European Defence Congress in Berlin, 9 December. Berlin: Europäischer Verteidigungskongress.

Schreer, Benjamin (2003): Die Transformation der U.S. Streitkräfte im Lichte des Irakkrieges. Berlin: SWP.

Schwiebert, Rainer (2004): Iraqi Freedom – Hat Network Centric Warfare die Feuertaufe bestanden? *Soldat und Technik*, February, 26–30.

Scott, Richard (2003): UK CEC Plots a Course for Network-Enabled Capability. *Jane's Navy International*, 17 December. http://search.janes.com (accessed 1 October 2010).

Scott, Richard W. (2001): Institutions and Organizations. Thousand Oaks, CA: Sage.

Scott, Richard W./Meyer, John W. (1991): The Organization of Societal Sectors: Propositions and Early Evidence. In: Powell, Walter W./DiMaggio, Paul J. (Eds.): The New Institutionalism in Organizational Analysis. Chicago, IL: University of Chicago Press, 108–140.

Scott, William B./Hughes, David (2003): Nascent Net-Centric War Gains Pentagon Toehold. *Aviation Week & Space Technology*, 4, 27 January, 50–53.

Shachtman, Noah (2007): How Technology Almost Lost the War: In: Iraq, the Critical Networks Are Social – Not Electronic. *Wired Magazine*, 27 November. http://www.wired.com/print/politics/security/magazine/15-12/ff_futurewar (accessed 25 October 2010).

Shalikashvili, John M. (1996): Joint Vision 2010. Washington, D.C.: Office of the Chairman of the Joint Chiefs of Staff.

Shamir, Eitan (2008): Military Culture and Mission Command (Auftragstaktik): A Case of Adoption and Adaptation. PhD thesis. London: King's College.

Shaw, Kevin/Ebbutt, Giles (2009): NEC in Action. In: British Ministry of Defence (Ed.): NEC – Understanding Network Enabled Capability. London: Ministry of Defence: 25–27.

Shelton, Henry H. (2000): Joint Vision 2020. Washington, D.C.: Office of the Chairman of the Joint Chiefs of Staff.

Simmons, Beth A./Dobbin, Frank/Garrett, Geoffrey (2006): Introduction: The International Diffusion of Liberalism. *International Organization*, 60: 4, 781–810.

Singer, P. W. (2009): Wired for War: The Robotics Revolution and Conflict in the Twenty-First Century. New York: Penguin Press.

Singhal, Arvind (2002): Speech introducing Professor Everett M. Rogers at the 47th Annual Research Lecturer, 24 April. Albuquerque: University of New Mexico.

Skinner, Tony (2006): Striving for NEC. *Jane's Defence Weekly*, 21 December. http://search.janes.com (accessed 23 April 2011).

Skinner, Tony (2008): UK Operations in Afghanistan 'Shaping' NEC Future. *Jane's Defence Weekly*, 18 April. http://search.janes.com (accessed 20 September 2010).

Sloan, Elinor C. (2008): Military Transformation and Modern Warfare: A Reference Handbook. Westport, CN: Praeger.

Smith, Eduard A. (2002): Effects Based Operations: Applying Network-Centric Warfare in Peace, Crisis and War. Washington, D.C.: CCRP Publication Series.

Smith, Rupert (2005): The Utility of Force: The Art of War in the Modern World. London: Penguin Books.

Snyder, Jack L. (1984): The Ideology of the Offensive: Military Decision-Making and the Disasters of 1914. Ithaca, N.Y.: Cornell University Press.

Soeters, Joseph/Fenema, Paul van/Beeres, Robertus Johannes Maria (2010): Managing Military Organisations: Theory and Practice. London – New York: Routledge.

Spiegel (2004): Peter Struck. *DER SPIEGEL*, 8: 164.

Standen, Iain (2010): Network Enabled Capability: A UK Perspective. *RUSI Defence Systems*, 13: 1, 78–81.

Stephens, Philip (2010): Realism Grabs Hold of British Foreign Policy. *Financial Times*. http://www.ft.com/cms/s/0/ea4783a8-8865-11df-aade-00144feabdc0.html (5 July 2010).

Storr, Jim (2009): The Failure of Digital Command and Information Systems. *RUSI Defence Systems*, 12: 2, 28–30.

Strassmann, Paul A. (2008): Does the 2009 Budget Support Network-Centric Missions? *SIGNAL Magazin (AFCEA)*, November. http://www.afcea.org/signal/articles/templates/Signal_Article_Template.asp?articleid=1748&zoneid=243 (accessed 21 June 2010).

Suchman, Mark C. (1995): Managing Legitimacy: Strategic and Institutional Approaches. *The Academy of Management Review*, 20: 3, 571–610.

Swan, Jacky/Newell, Sue/Robertson, Maxine (1999): Central Agencies in the Diffusion and Design of Technology: A Comparison of the UK and Sweden. *Organization Studies*, 20: 6, 905–931.

Swedish Ministry of Defence (2004): Our Future Defence – The Focus of Swedish Defence Policy 2005–2007. Stockholm: Swedish Government Bill, 2004/05: 5.

Talbot, David (2004): How Technology Failed in Iraq. *Technological Review*, November. http://www.technologyreview.com (accessed 10 July 2010).

Tangredi, Sam J. (2002): Assessing New Missions. In: Binnendijk, Hans (Ed.): Transforming America's Military. Center for Technology and National Security Policy. Washington, D.C.: National Defense University Press, 3–30.

Taylor, Trevor (2009): Jointery and the Emerging Defence Review (*RUSI Future Defence Review: Working Paper Series*, No. 4). London: RUSI.

Temperley, Martin (2009): Consistent Information Management and NEC. In: Ministry of Defence (Ed.): NEC – Understanding Network Enabled Capability. London: Ministry of Defence: 62–63.

Terriff, Terry (2006): Innovate or Die: Organizational Culture and the Origins of Maneuver Warfare in the United States Marine Corps. *Journal of Strategic Studies*, 29: 3, 475–503.

Terriff, Terry/Osinga, Frans P. B./Farrell, Theo (2010): A Transformation Gap? American Innovations and European Military Change. Stanford, CA: Stanford University Press.

Thelen, Kathleen (1999): Historical Institutionalism in Comparative Politics. *Annual Review of Political Science*, 2, 369–404.

Thelen, Kathleen (2004): How Institutions Evolve: The Political Economy of Skills in Germany, Britain, the United States, and Japan. Cambridge, MA – New York: Cambridge University Press.

Thomas, George M./Lauderdale, Pat (1988): State Authority and National Welfare Programs in the World System Context. *Sociological Forum*, 3: 3, 383–399.

Tobergte, Christian (2006): Möglichkeiten und Grenzen der Transformation amerikanischer Landstreitkräfte. Frankfurt am Main – Bonn: Report Verlag.

Toffler, Alvin/Toffler, Heidi (1993): War and Anti-War: Survival at the Dawn of the 21st Century. Boston, MA: Little, Brown & Co.

Tolbert, Pamela S. (1985): Institutional Environments and Resource Dependence: Sources of Administrative Structure in Institutions of Higher Education. *Administrative Science Quarterly*, 30: 1, 1–13.

Tolbert, Pamela S./Zucker, Lynne G. (1983): Institutional Sources of Change in the Formal Structure of Organizations: The Diffusion of Civil Service Reform, 1880–1935. *Administrative Science Quarterly*, 28: 1, 22–39.

U.S. Army (2001): U.S. Army White Paper: Concept for the Objective Force. Washington, D.C.: U.S. Army, Chief of Staff.

U.S. Department of Defense (2001): Network Centric Warfare. Report to Congress. Washington, D.C.: U.S. Department of Defense.

U.S. Department of Defense (2003): Transformation Planning Guidance. Washington, D.C.: U.S. Department of Defense.

U.S. Department of Defense (2008): Dictionary of Military and Associated Terms. Joint Publication (Joint Doctrine Division, Joint Staff), 1-02. Washington, D.C.: Joint Chiefs of Staff. http://www.dtic.mil/doctrine/jel/doddict/ (accessed 13 October 2008).

U.S. Joint Forces Command (USJFCOM) (2004): Joint Transformation Roadmap, 21 January. Norfolk: USJFCOM.

U.S. Joint Forces Command (USJFCOM) (2008): The Joint Operating Environment (JOE). Suffolk: USJFCOM.

U.S. National Defense Panel (1997): Transforming Defense. National Security in the 21st Century. Arlington, VA: National Defense Panel. http://www.dtic.mil/ndp/FullDoc2.pdf (accessed 9 October 2010).

Veblen, Thorstein (1899): The Theory of the Leisure Class: An Economic Study in the Evolution of Institutions. New York – London: Macmillan.

Vego, Milan (2007): The NCW Illusion. *Armed Forces Journal*, 1. http://www.armedforcesjournal.com/2007/01/2392378/ (accessed 20 September 2010).

Vennesson, Pascal/Breuer, Fabian/de Franco, Chiara/Schroeder, Ursula C. (2009): Is there a European Way of War? Role Conceptions, Organizational Frames, and the Utility of Force. *Armed Forces & Society*, 35: 4, 628–645.

Vickers, Michael G. (2001): Revolution Deferred: Kosovo and the Transformation of War. In: Bacevich, Andrew J./Cohen, Eliot A. (Eds.): War over Kosovo: Politics and Strategy in a Global Age. New York: Columbia University Press, 189–209.

Vickers, Michael J./Martinage, Robert C. (2004): The Revolution in War. Washington, D.C.: CSBA.

Wallace, William/Phillips, Christopher (2009): Reassessing the Special Relationship. *International Affairs*, 85: 2, 263–284.

Waters, Gary (2008): The Australian Defence Force and Network Centric Warfare. In: Waters, Gary/Ball, Desmond/Dudgeon, Ian (Eds.): Australia and Cyber-Warfare. Canberra: ANU E Press, 5–31.

Webster, Frederick E. (1971): Communication and Diffusion Processes in industrial Markets. *European Journal of Marketing*, 5: 4, 178–188.

Wehrtechnik, Studiengesellschaft der Deutschen Gesellschaft für (Ed.) (2003): Network Centric Capabilities und der Transformationsprozess. Bonn: DWT.

Weigley, Russell F. (1977): The American Way of War: A History of United States Military Strategy and Policy. Bloomington, IN: Indiana University Press.

Weinraub, Bernard/Shanker, Thom (2003): Rumsfeld's Design for War Criticized on the Battlefield. *The New York Times,* 31 March, 7.

Weiss, Robert St. (1994): Learning from Strangers: The Art and Method of Qualitative Interview Studies. New York: Free Press.

Wendt, Alexander (1999): Social Theory of International Politics. Cambridge – New York: Cambridge University Press.

Whittle, Richard (2001): War Speeds up Pace of Change for Military. *The Dallas Morning News,* 5 December, 16.

Wiesner, Ina (2010): NCO in Germany. Still a Long Way to Go. *RUSI Defence Systems*, 13: 1, 82–84.

Williams, John/Foley, Gerald (2003): Lessons for NEC from Previous Information Systems Initiatives. *Journal of Defence Science*, 8: 3, 168–171.

Williams, Rudi (2003): 'Horizontal Fusion' Makes Troops Less Vulnerable, More Lethal. Washington, D.C.: Department of Defense News, American Forces Press Service. http://www.defense.gov/News/NewsArticle.aspx?ID=28412 (accessed 28 September 2010).

Wilson, Clay (2007): Network Centric Warfare. Background and Oversight Issues for Congress. Updated 15 March 2007. Document No RL32411. Washington, D.C.: Congressional Research Service.

Wooten, Melissa/Hoffman, Andrew J. (2008): Organizational Fields: Past, Present and Future. In: Greenwood, Royston/Oliver, Christine/Sahlin, Kerstin/Suddaby, Roy (Eds.): The SAGE Handbook of Organizational Institutionalism. Los Angeles, CA – London: SAGE, 130–147.

Wright, Donald P. (2010): A Different Kind of War: The United States Army in Operation Enduring Freedom (OEF), October 2001 – September 2005. Fort Leavenworth, KS: Combat Studies Institute Press, U.S. Army Combined Arms Center.

Young, Thomas-Durell (1994): Trends in German Defense Policy: The Defense Policy Guidelines and the Centralization of Operational Control. London: The Royal Institute of International Affairs.

Young, Thomas-Durell (1996): German National Command Structures after Unification: A New German General Staff? *Armed Forces & Society*, 22: 3, 379–417.